Reviews for Other Books by Yoshihito Isogawa

THE LEGO TECHNIC IDEA BOOK SERIES:

"These are an invaluable set of books to have as a reference to build mechanisms."

—JOE MENO, *BrickJournal*

"These are excellent books showing a lot of great ideas for LEGO mechanisms. Even if you're an experienced builder, there are surely some ideas in here you've never seen."

—BILL WARD, Brickpile

"What I like about these cool little models is that they can be used to teach various science concepts such as gearing, Newton's Laws, and potential & kinetic energy—to name a few."

—THE ROBOTIC REALM

"For anyone who loves LEGO, prototypes in LEGO, or loves mechanical assemblies, these books are definitely required viewing, and we're not sure how we lived without them for so long."

—LENORE EDMAN, Evil Mad Scientist Laboratories

THE LEGO MINDSTORMS EV3 IDEA BOOK:

"Minimalistic but highly informative…encouraging young engineers to apply problem solving and creativity to the endless combinations of mechanics."

—*BOOKLIST*

THE LEGO BOOST IDEA BOOK:

"Yoshihito's books are among the most useful I own."

—BRICKSET

"For experienced builders and BOOST programmers, there is plenty of content to inspire creative new builds and stretch their imagination with original builds."

—GEEKDAD

LEGO® Technic™
Non-Electric Models
Simple Machines

YOSHIHITO ISOGAWA

**no starch
press**

LEGO TECHNIC NON-ELECTRIC MODELS: SIMPLE MACHINES. Copyright © 2021 by Yoshihito Isogawa.

Printed in the United States of America

First printing

25 24 23 22 21   1 2 3 4 5 6 7 8 9

ISBN-13: 978-1-7185-0120-1 (print)
ISBN-13: 978-1-7185-0121-8 (ebook)

Publisher: William Pollock
Executive Editor: Barbara Yien
Production Manager: Rachel Monaghan
Production Editors: Kassie Andreadis and Paula Williamson
Cover Design: Monica Kamsvaag
Photographer: Yoshihito Isogawa
Author Photo: Sumiko Hirano
Technical Reviewer: Sumiko Hirano
Developmental Editor: Nathan Heidelberger
Copyeditor: Rachel Monaghan
Compositor: Maureen Forys, Happenstance Type-O-Rama
Proofreader: Emelie Battaglia

For information on book distributors or translations, please contact No Starch Press, Inc. directly:

No Starch Press, Inc.
245 8th Street, San Francisco, CA 94103
phone: 1-415-863-9900; info@nostarch.com
www.nostarch.com

Library of Congress Cataloging-in-Publication Data

Names: Isogawa, Yoshihito, 1962– author.
Title: LEGO Technic non-electric models / Yoshihito Isogawa.
Description: San Francisco : No Starch Press, [2021] | Contents: volume 1.
   Simple machines — volume 2. Clever contraptions
Identifiers: LCCN 2021008259 (print) | LCCN 2021008260 (ebook) | ISBN
   9781718501201 (v. 1 ; paperback) | ISBN 9781718501706 (v. 2 ; paperback)
   | ISBN 9781718501218 (v. 1 ; ebook) | ISBN 9781718501713 (v. 2 ; ebook)
Subjects: LCSH: Machinery—Models—Juvenile literature. | LEGO
   toys—Juvenile literature.
Classification: LCC TJ248 .I8635 2021 (print) | LCC TJ248 (ebook) | DDC
   629.22/1—dc23
LC record available at https://lccn.loc.gov/2021008259
LC ebook record available at https://lccn.loc.gov/2021008260

[BB]

Contents

PART 1   Basic Mechanisms

PART 2 Moving Vehicles

Introduction

This is an idea book, offering more than 100 models you can build with LEGO Technic parts. The book focuses on models that you can build and operate without using electric elements such as motors. With these fun non-electric projects, you'll learn mechanical engineering principles through hands-on play.

How to Use This Book

This book doesn't include step-by-step building instructions. Instead, you'll find photographs of each model taken from various angles, plus a list of the parts you'll need for the model. Look at the photographs closely and try to reproduce the models. Building in this way is like putting together a puzzle.

You don't have to build the models in order. Flip through the pages and try making the ones you find most interesting. However, you might want to start with relatively simple models until you get used to the process.

As you build the models, pay attention to their movements and try to figure out why they work the way they do. This will help you develop your building skills. The next step is to adapt the ideas in this book to create your own original projects. If you need inspiration, refer to the hints that accompany many of the models. You can also combine ideas from multiple projects. Rework, reinforce, and decorate the models—your creativity has no limits!

LEGO Parts

At the end of the book, there's a list of all the parts you'll need to complete the projects. Most of the parts are common and easily obtainable. However, if you're missing a few, think of ways to swap in parts that you already own.

The images in this book feature parts in colors that make it easier for you to see the individual bricks' shapes. But you don't have to use the colors I've chosen; use any colors you want to make the projects your own.

This book's projects were designed to be built specifically with LEGO bricks and LEGO Technic parts made by the LEGO Group. I can't guarantee that the projects will work properly if made with non-LEGO products, which may have problems with accuracy or durability.

Further Reading

This book is part of a set. For even more fun non-electric projects, check out *LEGO Technic Non-Electric Models: Clever Contraptions*. It features a variety of interesting models, including drawing devices, spinning tops, measuring tools, and stands for a smartphone or tablet.

If you'd also like to learn about mechanisms that use electric motors, check out two of my other books: *The LEGO Power Functions Idea Book, Vol. 1* and *The LEGO Power Functions Idea Book, Vol. 2* (both from No Starch Press, 2015).

Acknowledgments

LDraw data and the LPub application were used to create the illustrations in this book. I would like to thank those involved in the development of those programs.

Warm-up

You won't find step-by-step building instructions in this book. Instead, you'll use photographs taken from various angles to try to reproduce the model shown. Building in this way is like putting together a puzzle. You'll soon get the hang of this process and learn to enjoy it! Let's practice first.

#1

This is the number of the model.

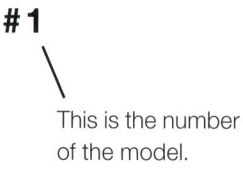

All the parts you need for this model are shown below. Find them among the parts you own, and start building!

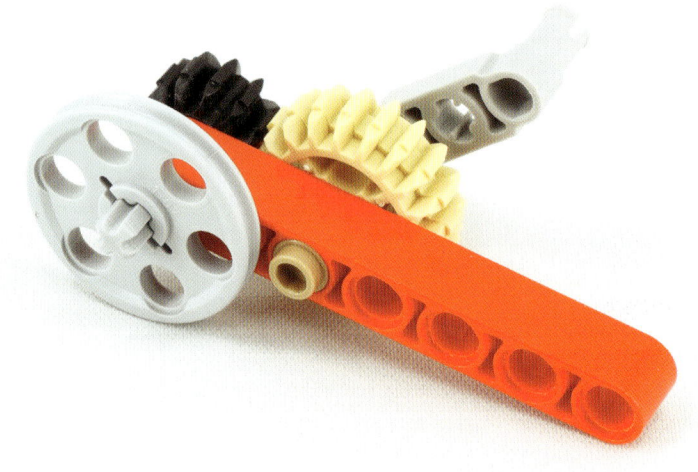

After gathering your parts, try building the model using the photos on this page and the next. To work faster, put your model in the same position as the one in the photos, and keep comparing them as you build.

This is the hint icon, which suggests other ways of building the project. Try to create your own unique and fun models using these tips. Please note that the parts used in the hint are not included in the parts list for each project or at the end of the book.

PART 1
Basic Mechanisms

The most basic of basics

Length

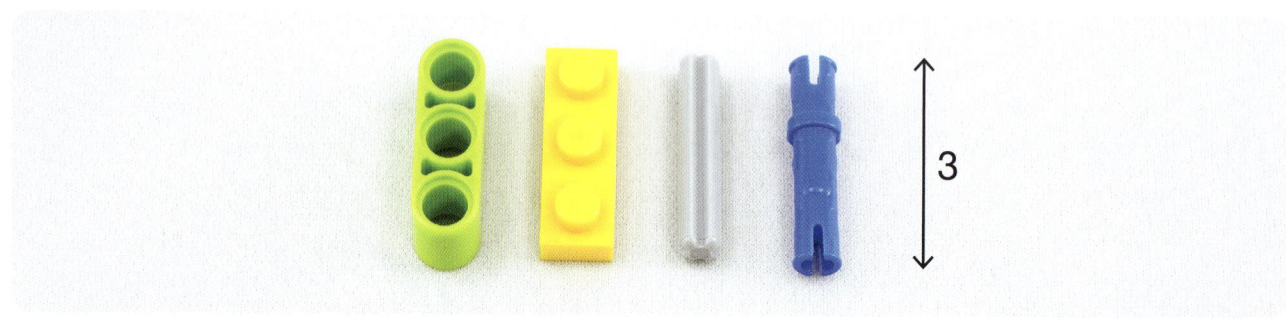

3

5
7
9
11
13
15

2
3
4
5
6
7
8
9
10
12

Number of teeth

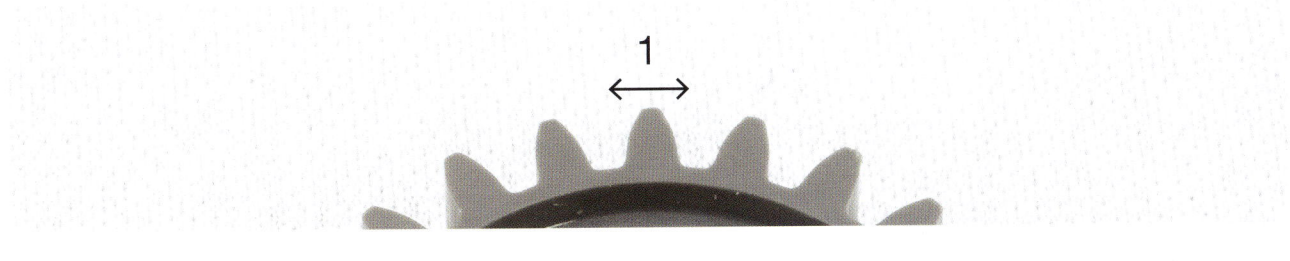

1

8 12 16 20 24 28 36 40

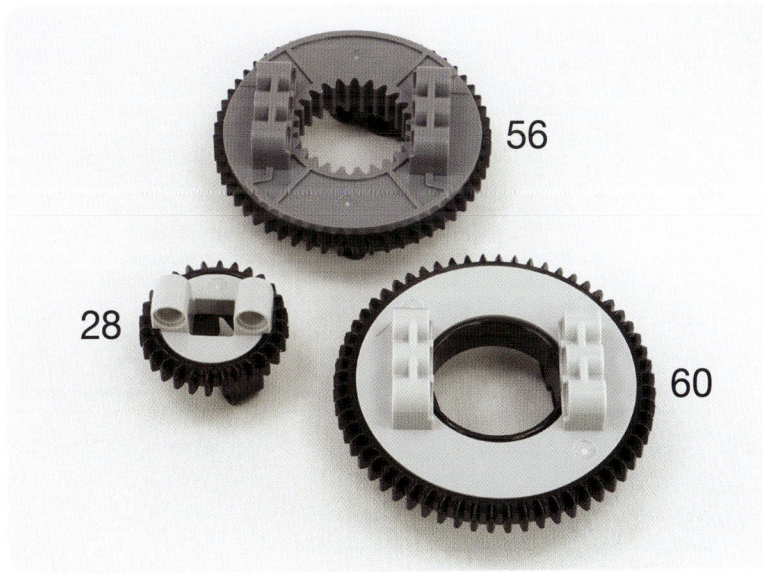

56

28

60

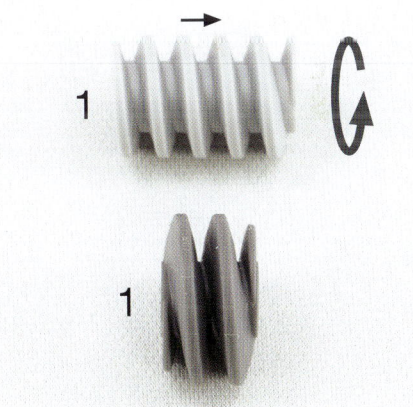

LEGO worm gears move one tooth during one rotation. So you can think of them as 1-tooth gears.

1

1

Connectors with and without friction

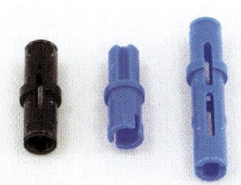

Friction connectors are used to secure one part to another.

Non-friction connectors are used as fulcrums for moving mechanisms.

Transmitting rotation with gears

Two meshing gears turn in opposite directions.

Turn this handle.

When two gears with the same number of teeth are used to transmit rotation, there's no change in the speed or power.

SAME SPEED **SAME POWER**

#2

When rotation is transmitted from a large gear to a small gear, the speed increases and the power decreases.

SPEED POWER
UP DOWN
3:1 (24:8)

This number is the *gear ratio*, a ratio of the number of teeth in the transmitting gear compared to the receiving gear.

#3

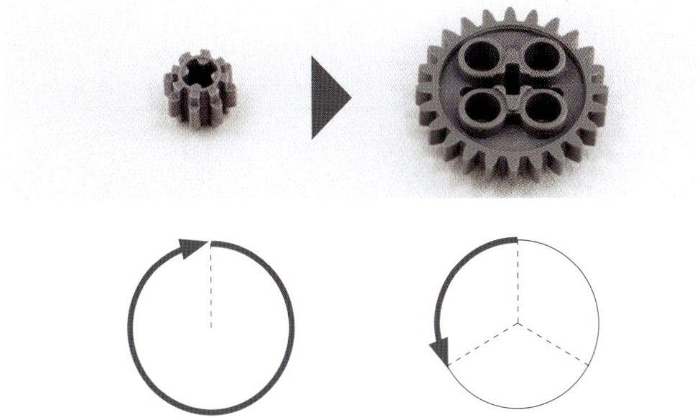

When rotation is transmitted from a small gear to a large gear, the speed decreases and the power increases.

SLOW POWER
DOWN UP
1:3 (8:24)

This model has a gear ratio of one to three.

Handle

SLOW POWER
DOWN UP
3:5 (12:20)

💡 You'll use a handle to move a lot of the models in this book. If you don't have this part, or if you want to express your own personality with the model, you can use other parts for the handle.

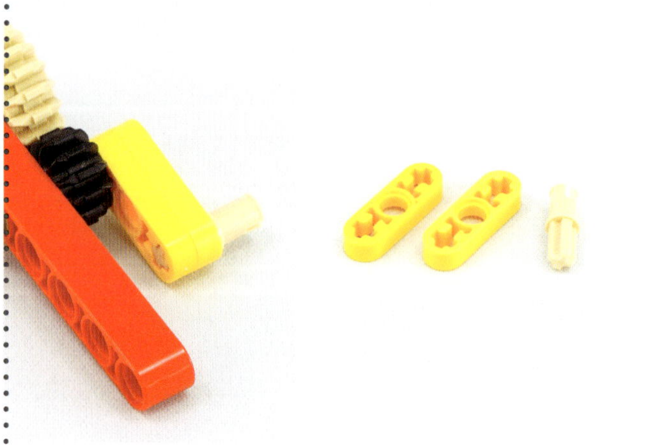

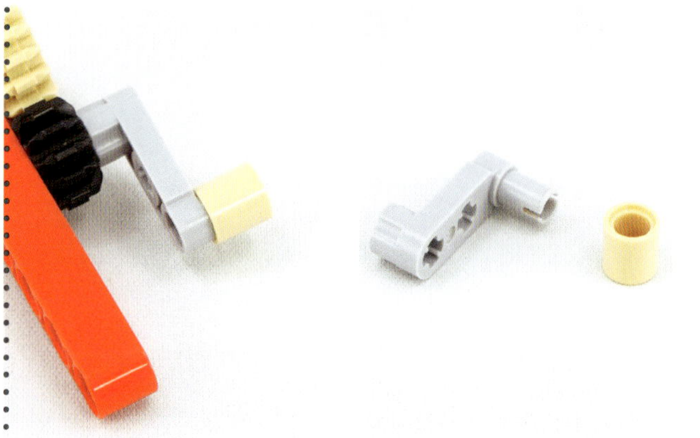

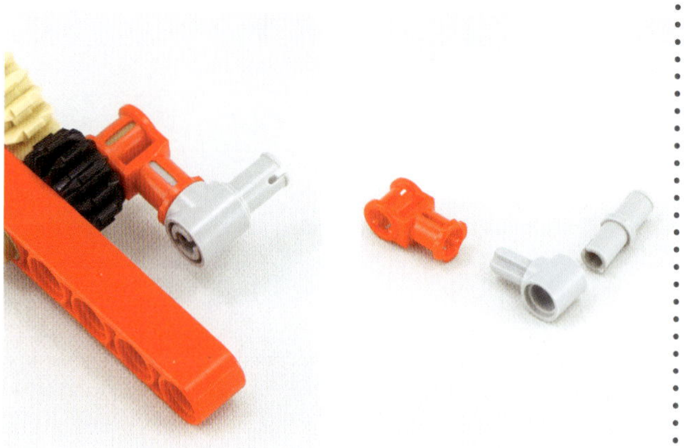

#5

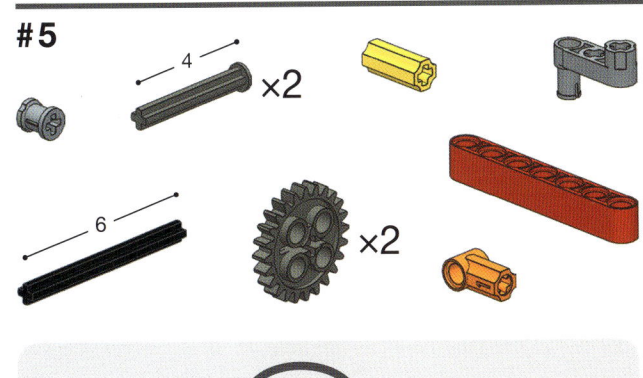

SAME SPEED   SAME POWER

#6

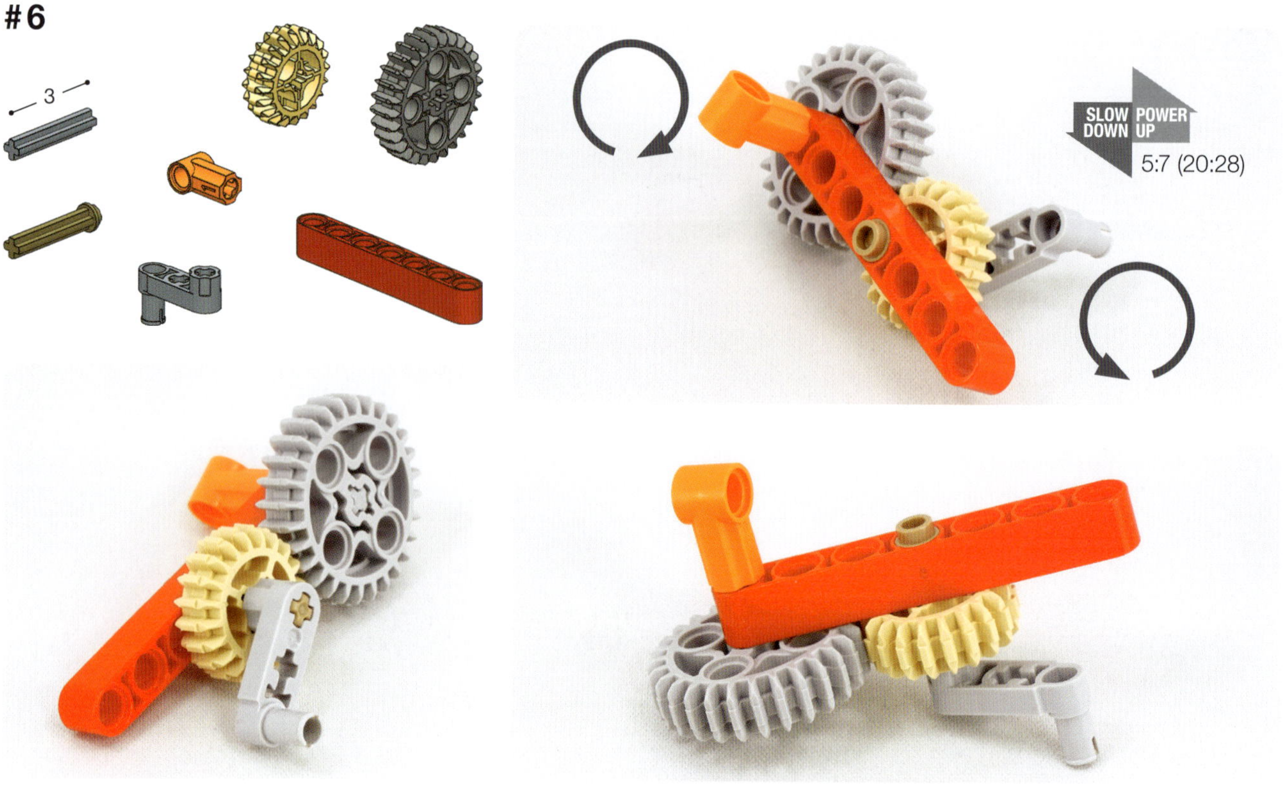
3
SLOW DOWN
POWER UP
5:7 (20:28)

#7

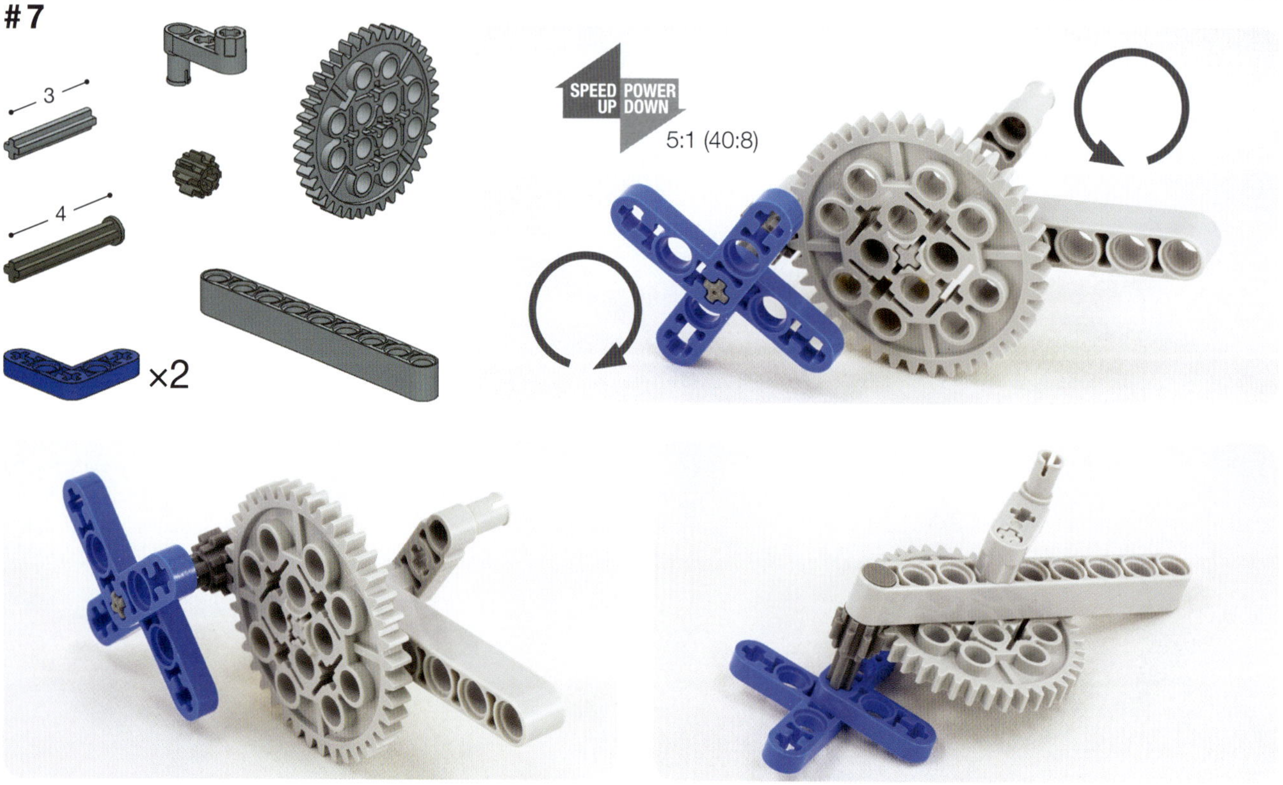
3
4
×2
SPEED UP
POWER DOWN
5:1 (40:8)

#8

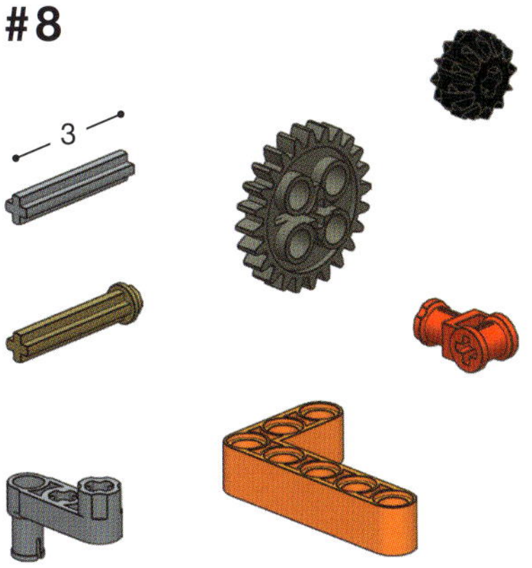

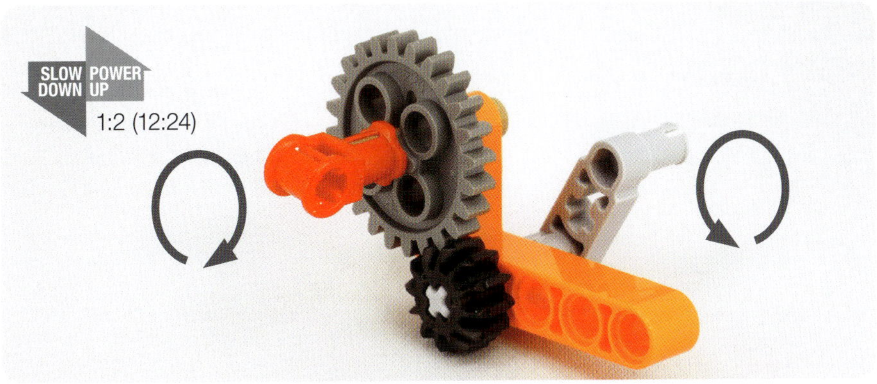

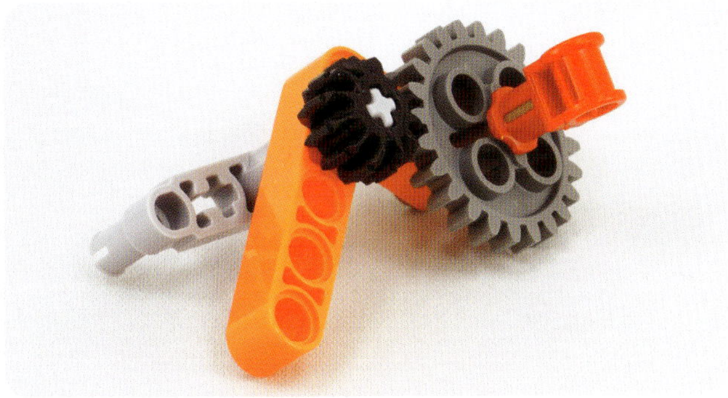

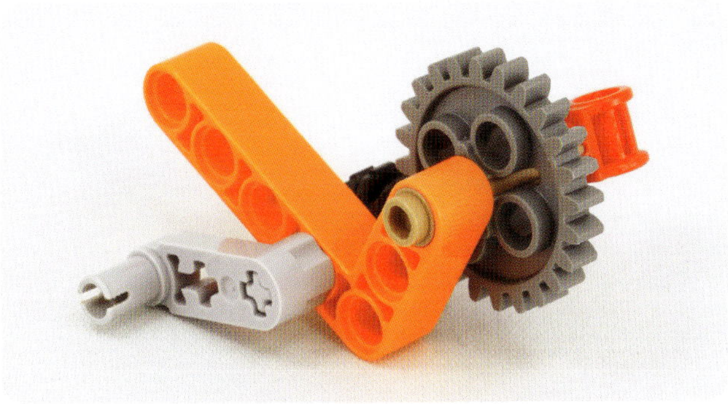

SLOW POWER
DOWN UP
1:2 (12:24)

#9

SPEED POWER
UP DOWN
6:5 (24:20)

#10

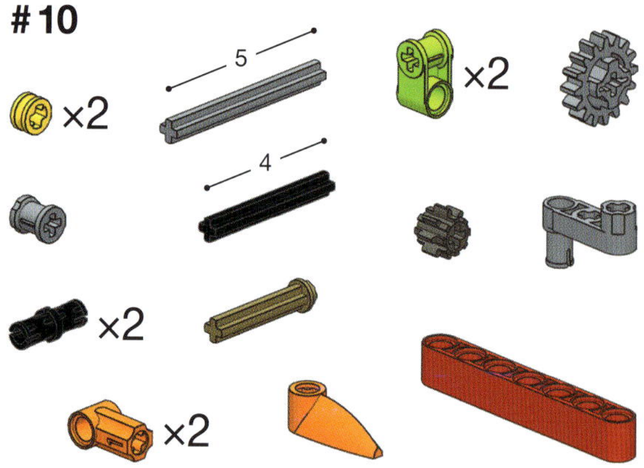

×2 —5— ×2

×2

×2 —4—

SLOW DOWN POWER UP
1:2 (8:16)

—5—

#11

×4 ×2

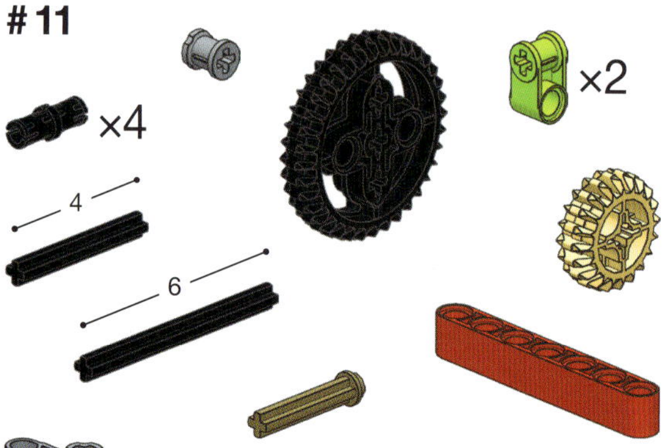

—4—

—6—

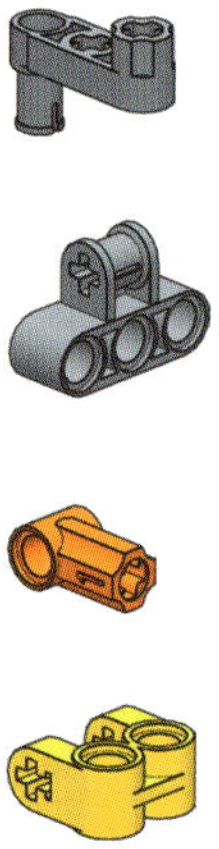

SLOW DOWN POWER UP
5:9 (20:36)

10

#12

×2 ×2

— 3 — ×2

SPEED POWER
UP DOWN
3:1 (24:8)

#13

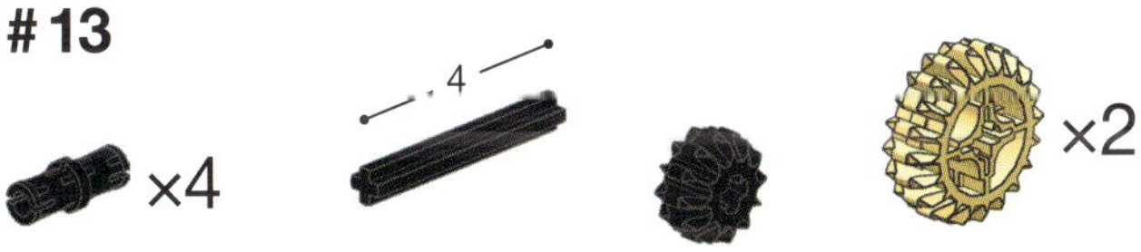

×4 . 4 ×2

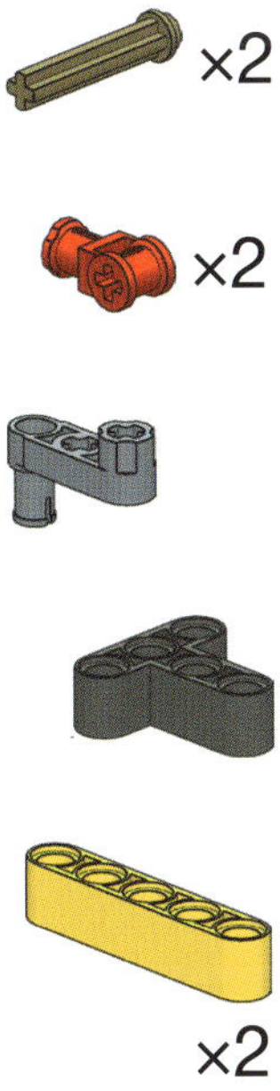

×2

×2

×2

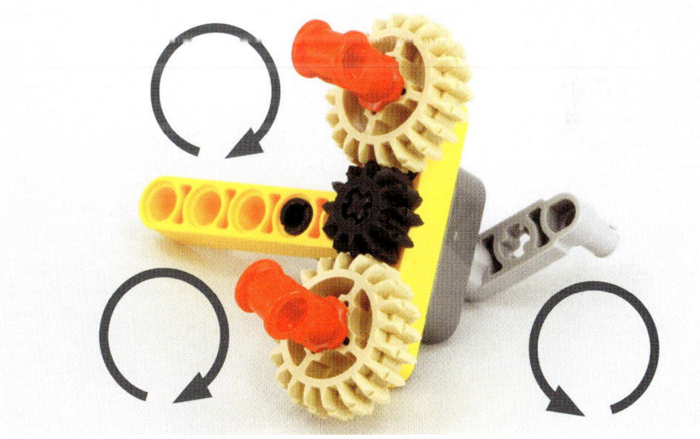

SLOW POWER
DOWN UP
3:5 (12:20)

×2

14

SLOW DOWN **POWER UP**

1:3 (8:24)

When gears are aligned consecutively, only the first and last gears affect speed and power, not the ones in between. That's because the gears in between are passing only one tooth's worth of rotation to the neighboring gears.

24 — 8 — 8 — 24 — 8 — 8

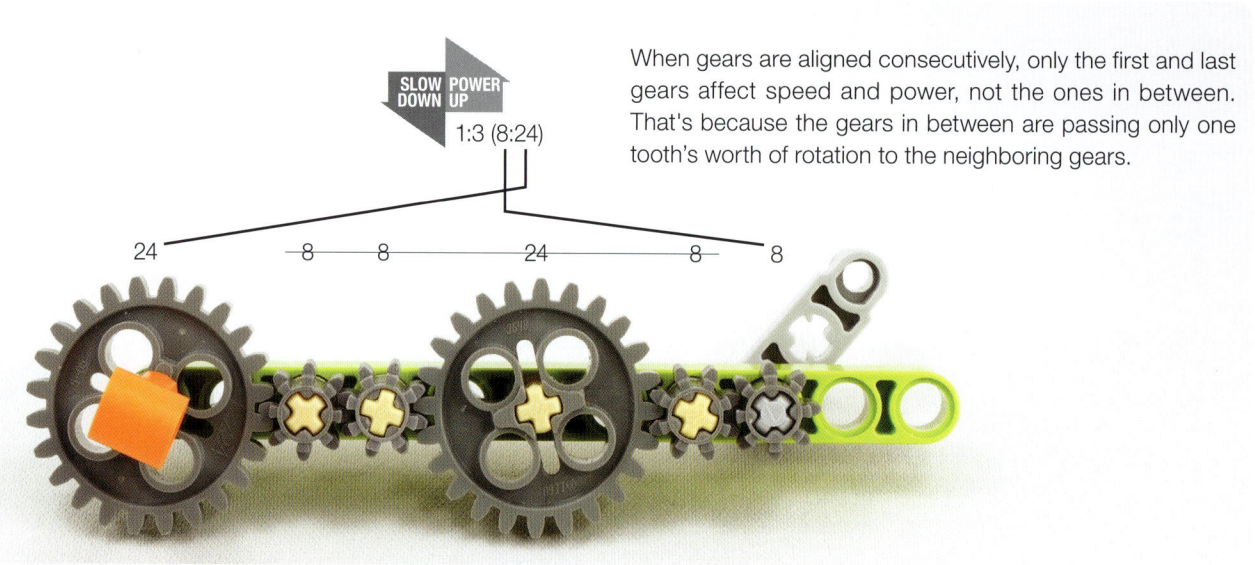

12

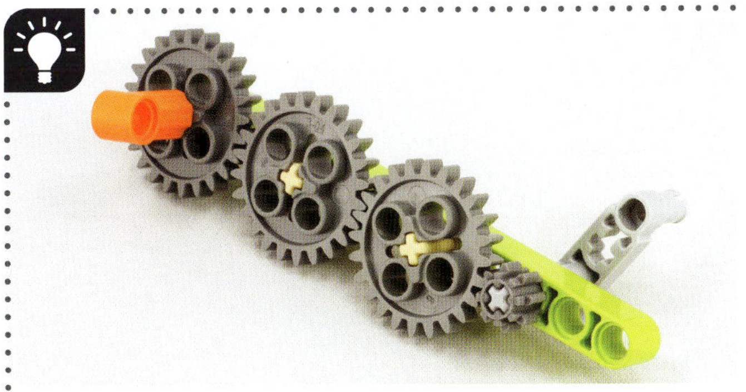

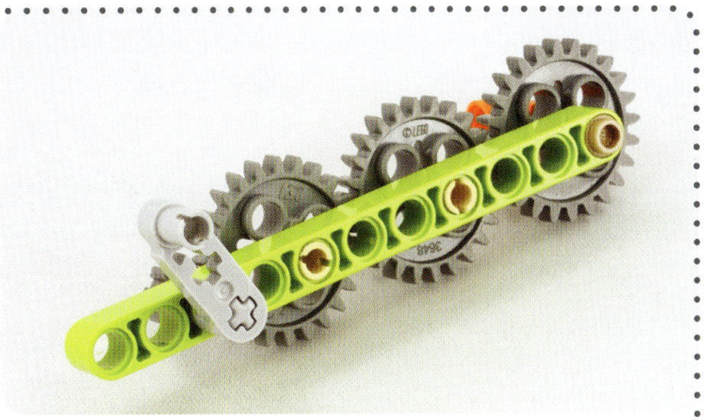

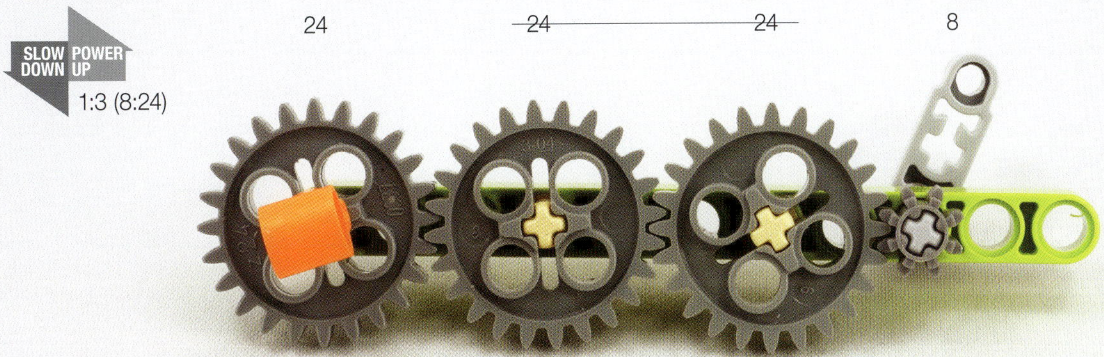
SLOW DOWN
POWER UP
1:3 (8:24)
24
24
24
8

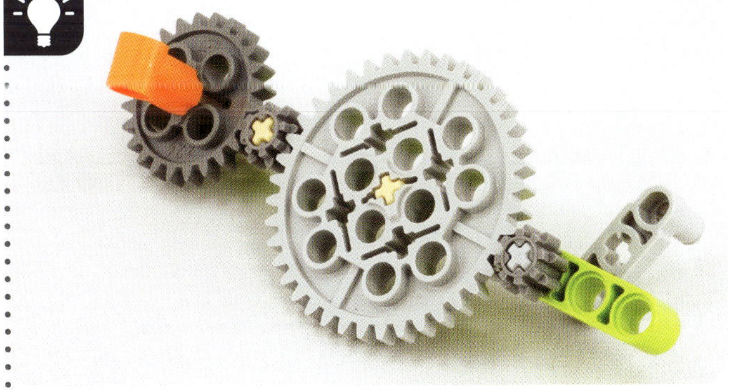

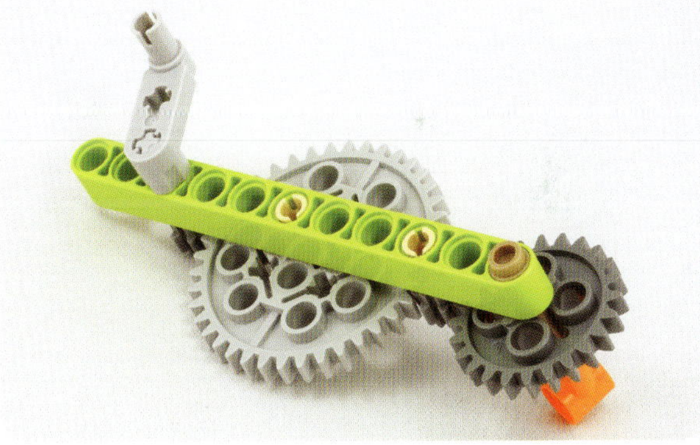

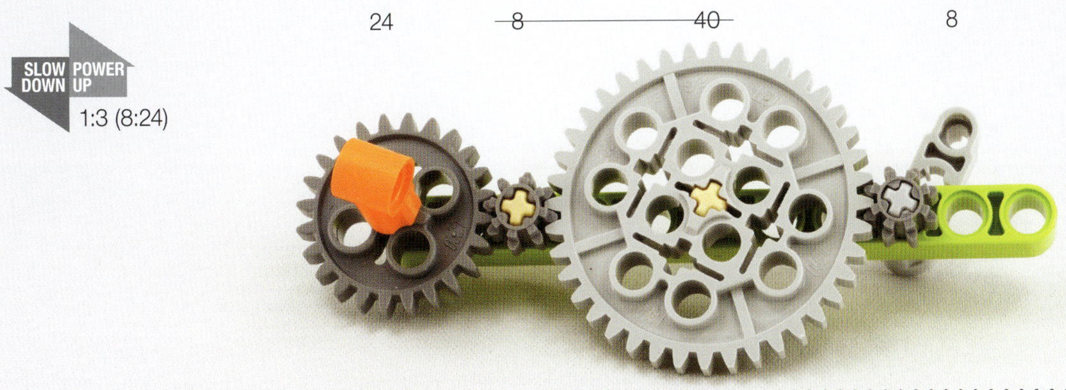
SLOW DOWN
POWER UP
1:3 (8:24)
24
8
40
8

#15

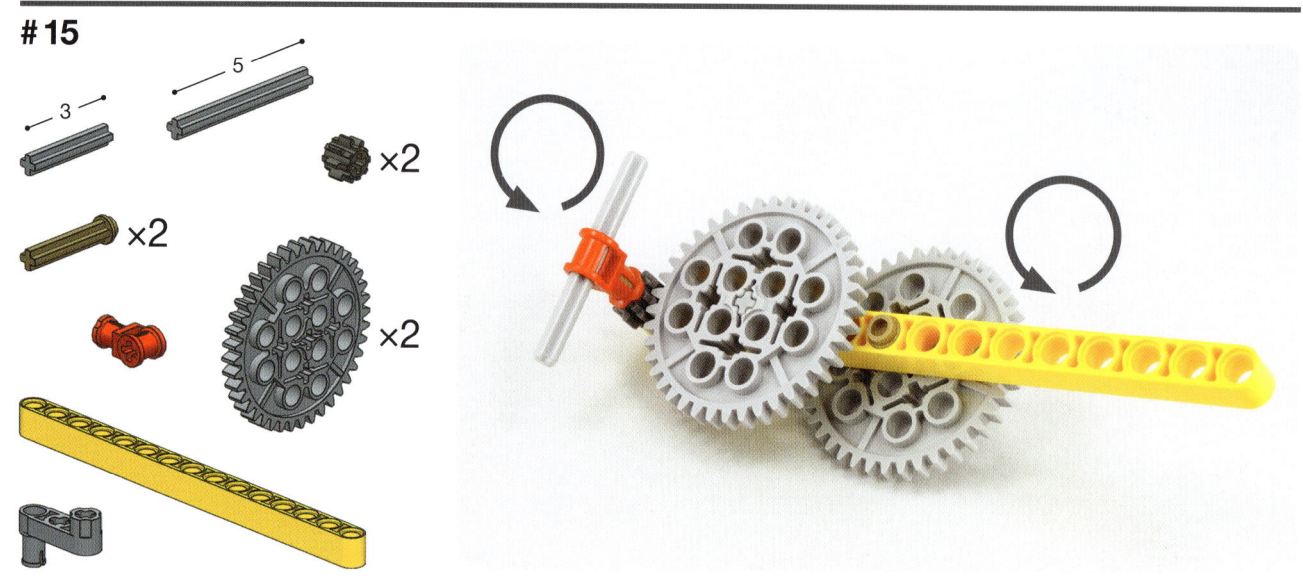

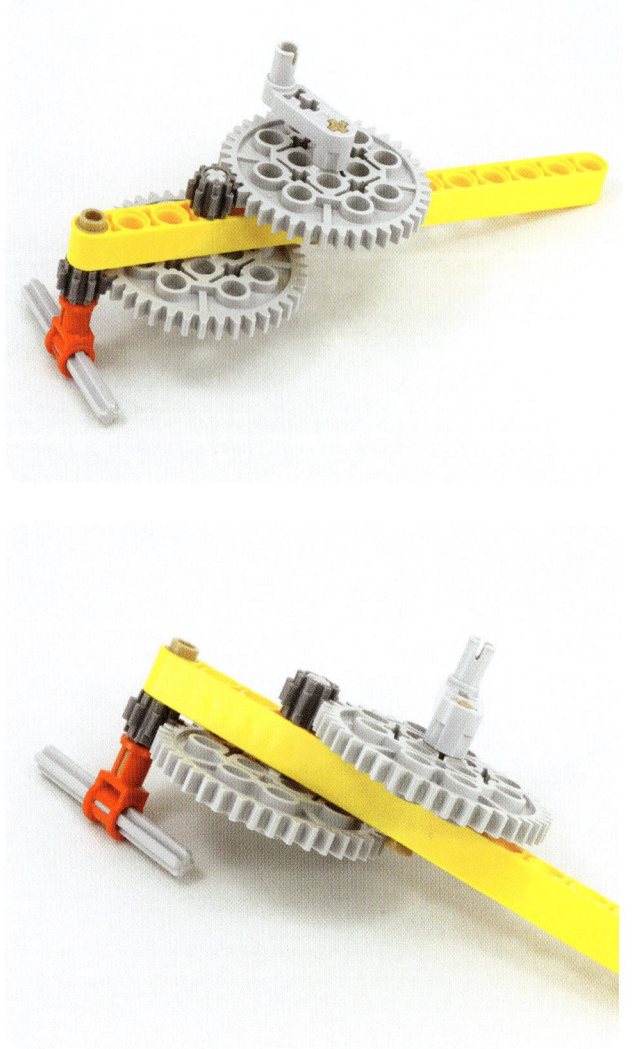

Combining gears in this way allows for even greater changes in speed and power. In this example, by repeating the 5-fold speed-up twice, you get a 5 × 5 = 25-fold speed-up.

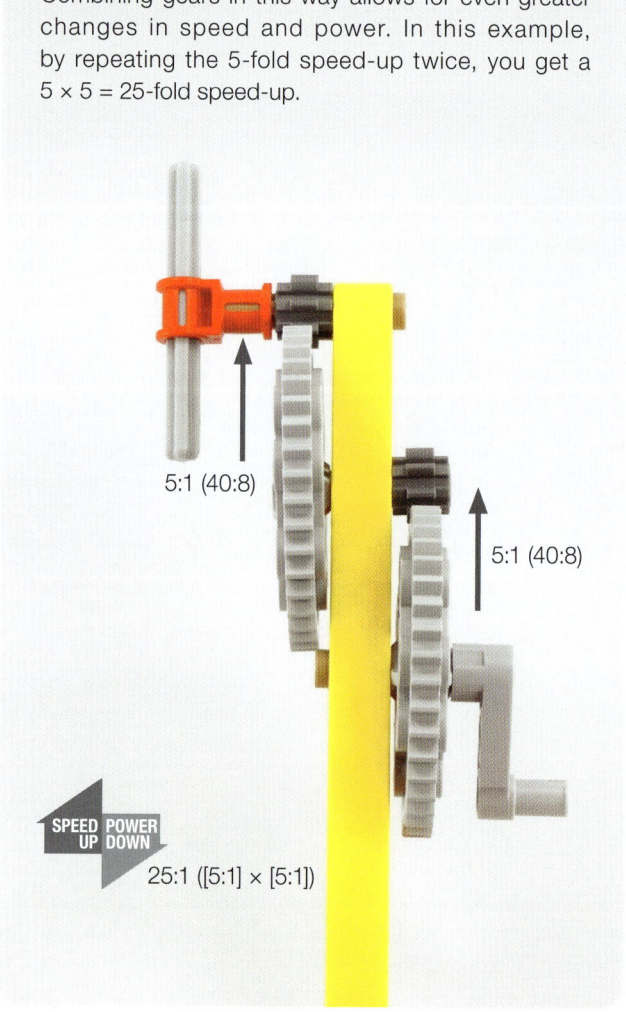

5:1 (40:8)

5:1 (40:8)

SPEED UP / POWER DOWN

25:1 ([5:1] × [5:1])

#16

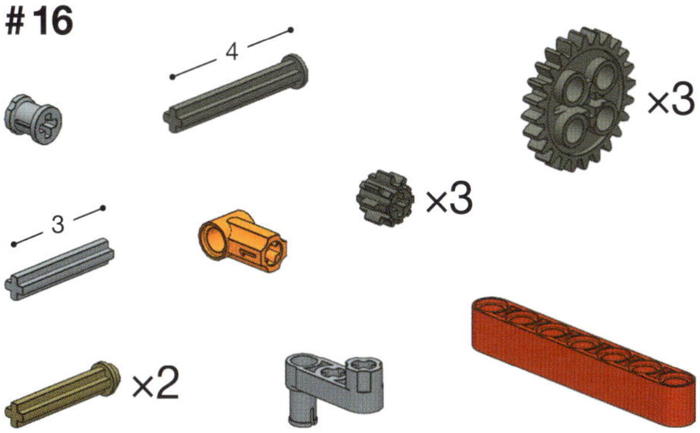
4
×3
×3
3
×2

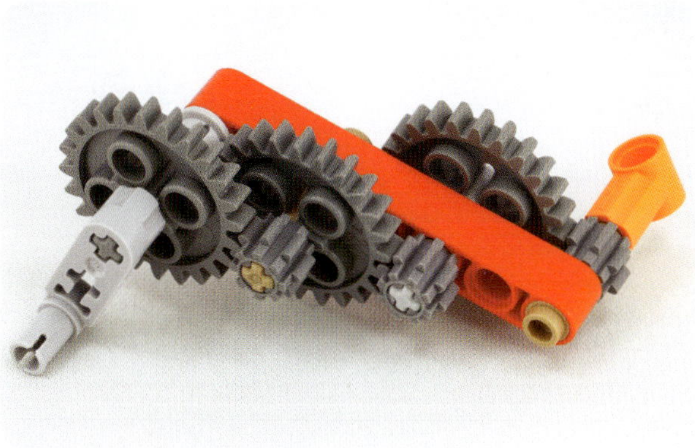

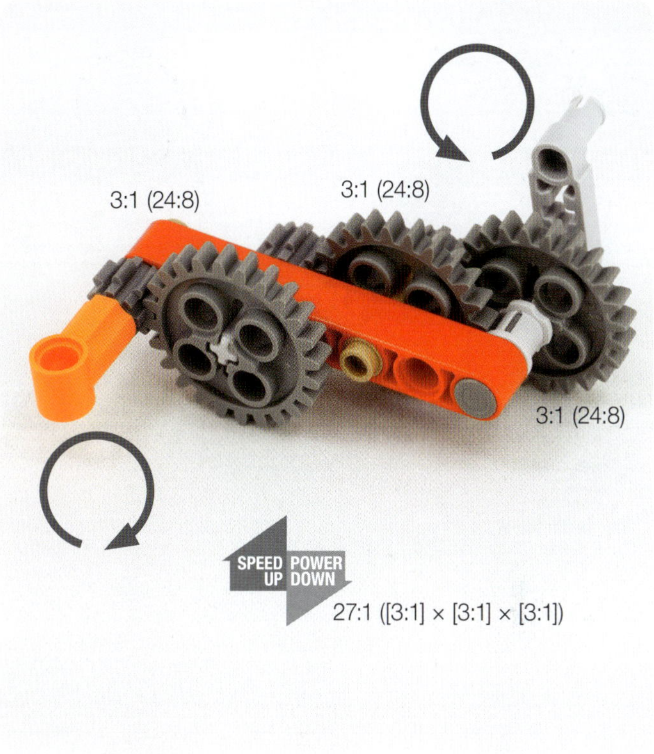
3:1 (24:8)
3:1 (24:8)
3:1 (24:8)
SPEED UP
POWER DOWN
27:1 ([3:1] × [3:1] × [3:1])

#17

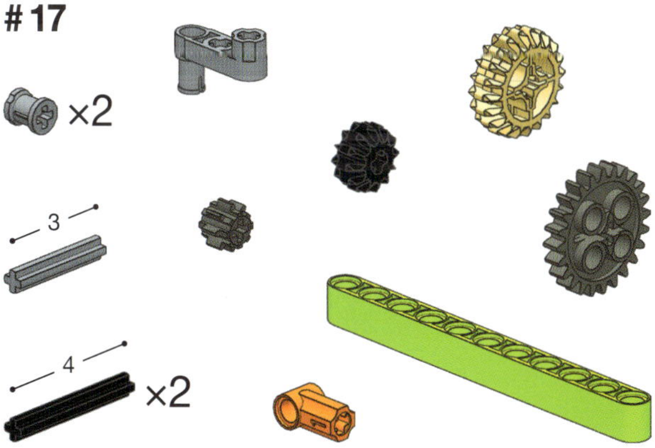
×2
3
4
×2

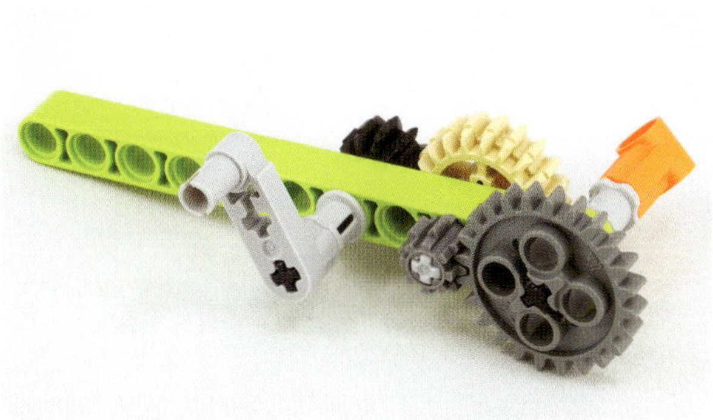

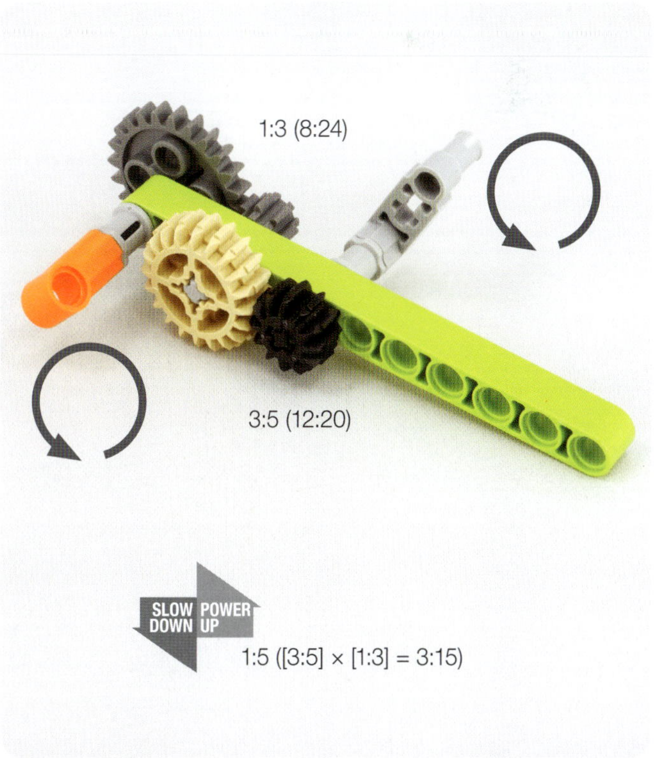
1:3 (8:24)
3:5 (12:20)
SLOW DOWN
POWER UP
1:5 ([3:5] × [1:3] = 3:15)

Changing the angle of rotation

 # 18

×2

×2

SAME SPEED SAME POWER

#19

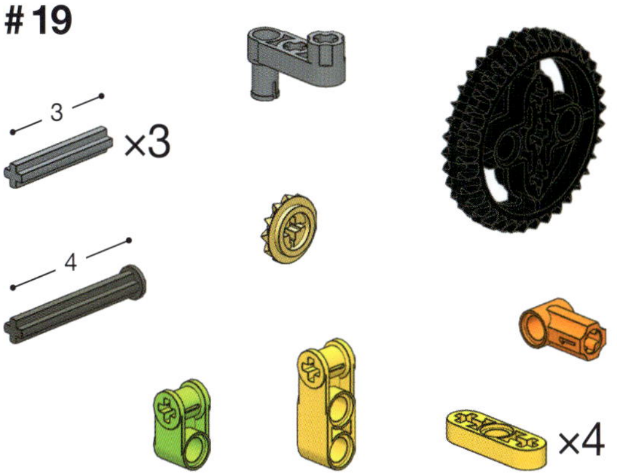

×3

×4

SLOW POWER
DOWN UP
1:3 (12:36)

#20

×2

×2

SAME SAME
SPEED POWER

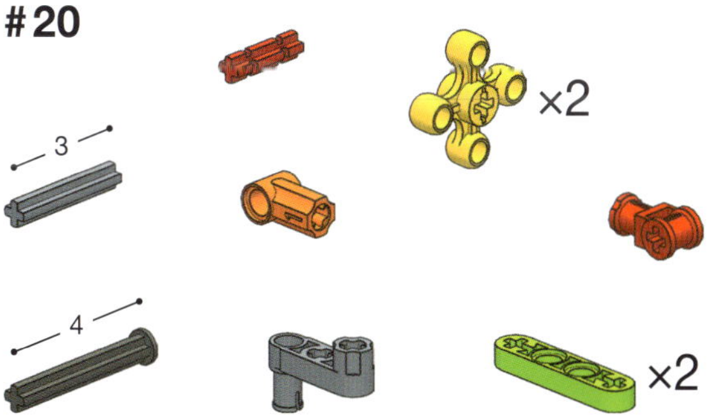

#21

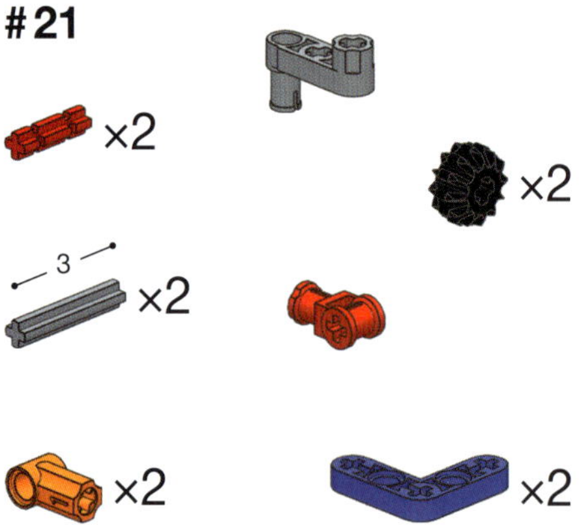

×2

×2

3
×2

×2

×2

×2

SAME SPEED | SAME POWER

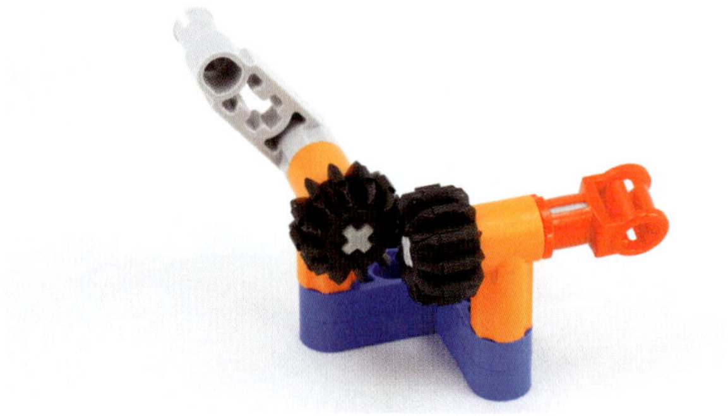

#22

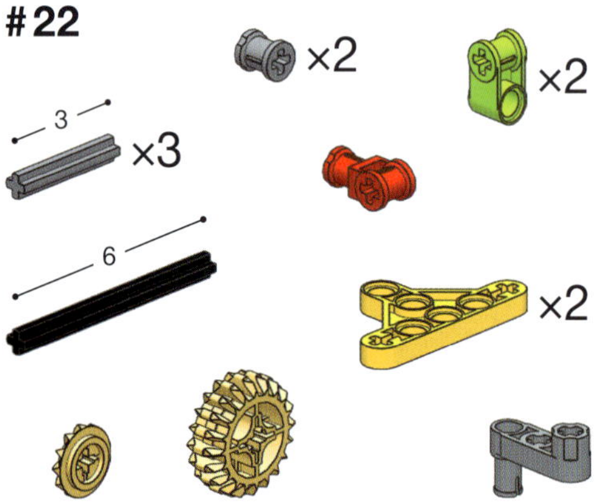

×2

×2

3
×3

×2

6

×2

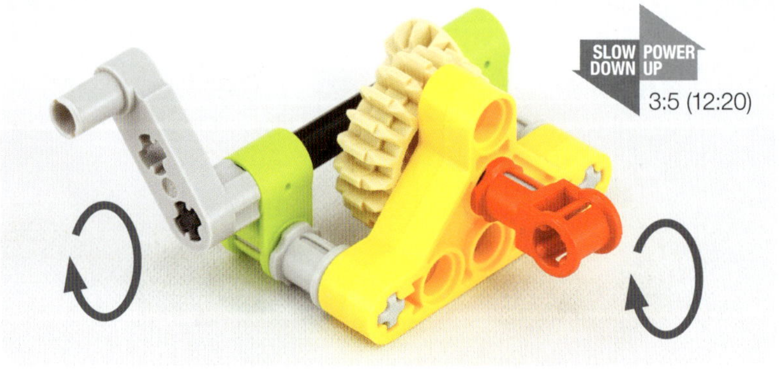

SLOW DOWN | POWER UP
3:5 (12:20)

#23

×2

×2

**SLOW
DOWN** | **POWER
UP**
2:3 (16:24)

#24

×3

×2

×2

**SLOW
DOWN** | **POWER
UP**
1:3 (8:24)

#25

×4

5

7

×2

×2

×2

×2

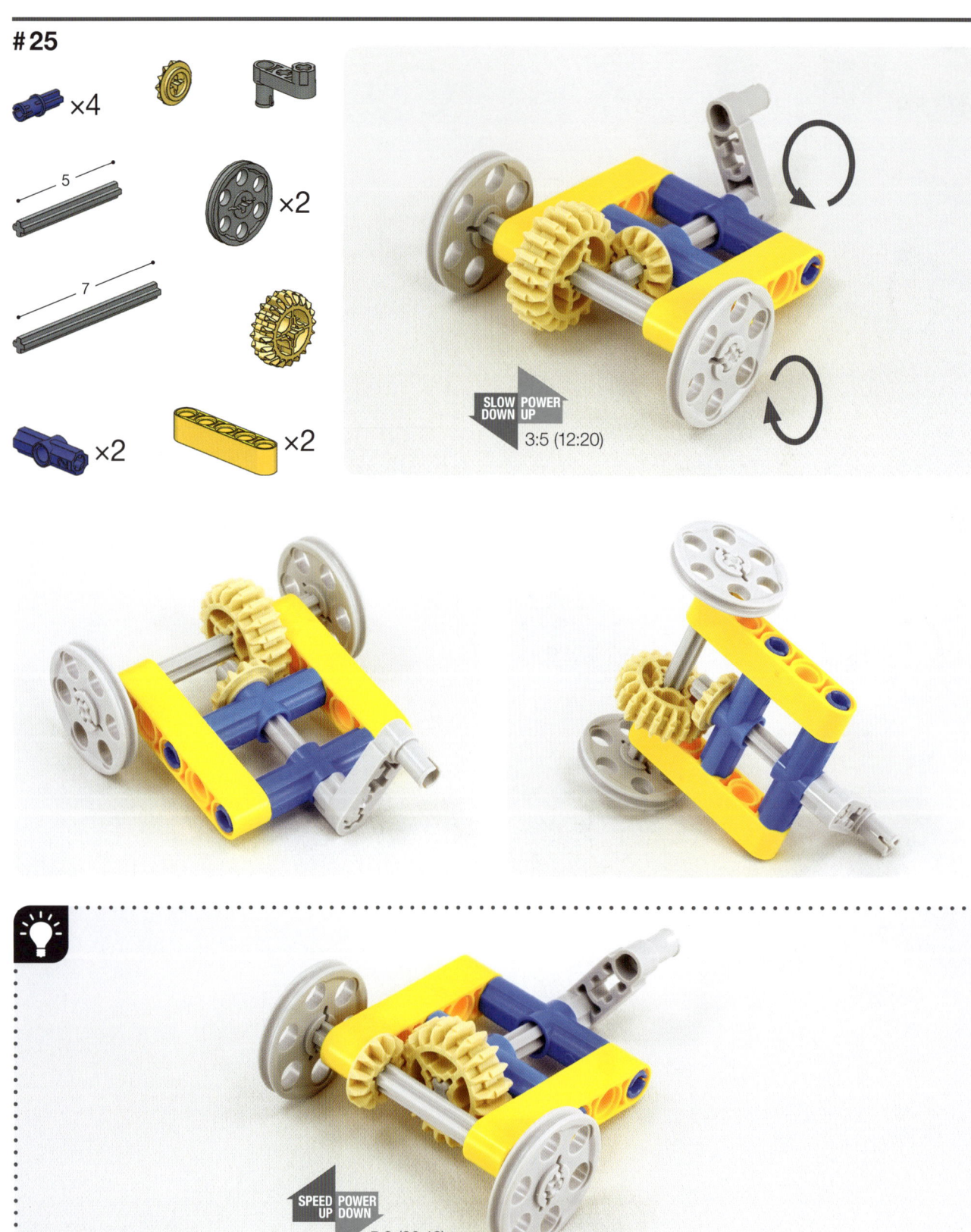

SLOW POWER
DOWN UP
3:5 (12:20)

SPEED POWER
UP DOWN
5:3 (20:12)

#26

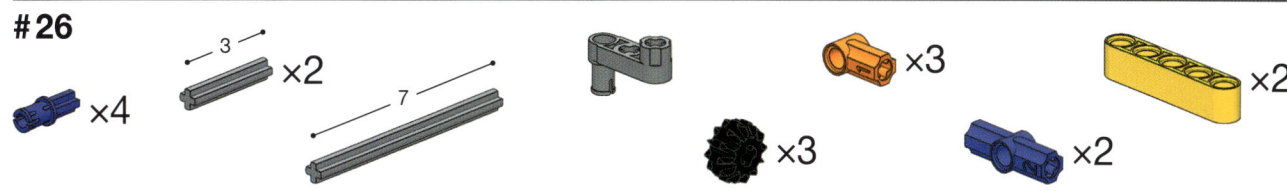

×4 ×2 —3— —7— ×3 ×3 ×2 ×2

SAME SAME
SPEED POWER

SAME SAME
SPEED POWER

SAME SAME
SPEED POWER

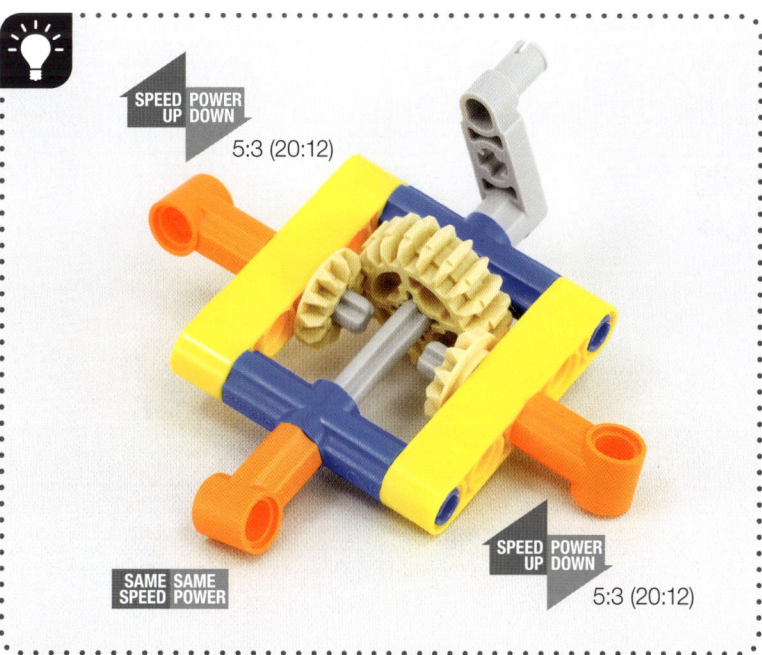

SPEED POWER
UP DOWN
5:3 (20:12)

SAME SAME
SPEED POWER

SPEED POWER
UP DOWN
5:3 (20:12)

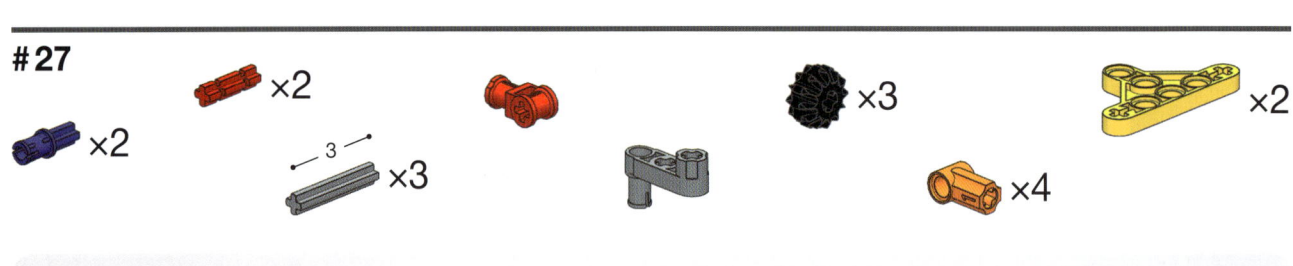

×2 ×2 ×3 ×2

×2 — 3 — ×3 ×4

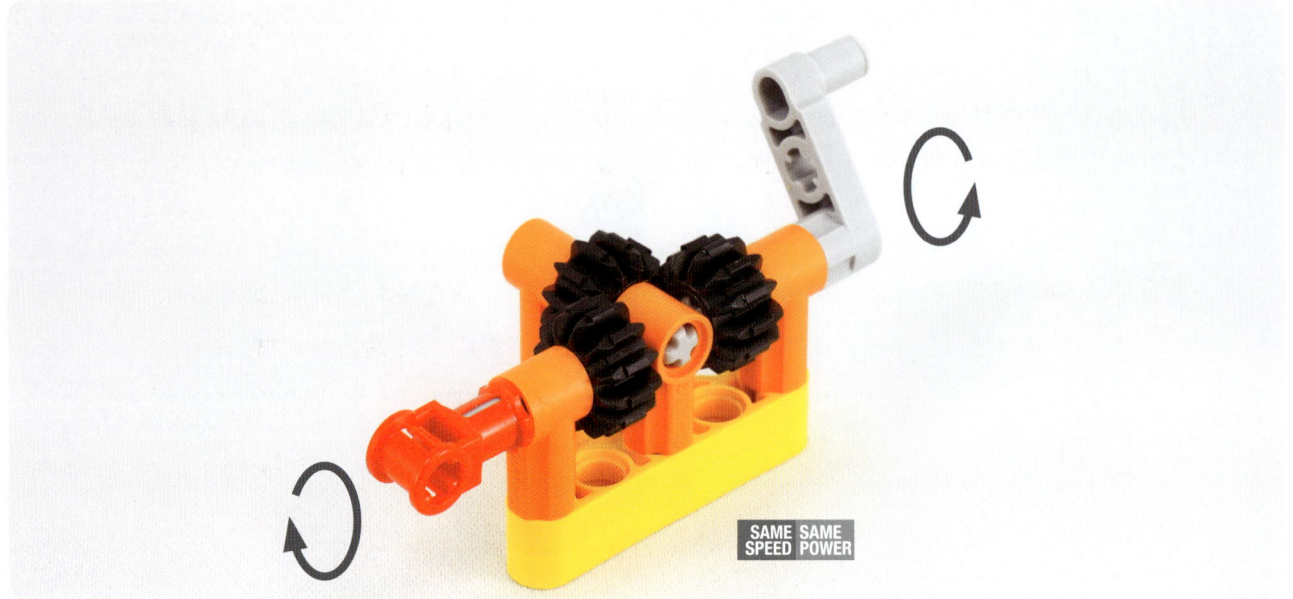

SAME SPEED SAME POWER

×4

×2

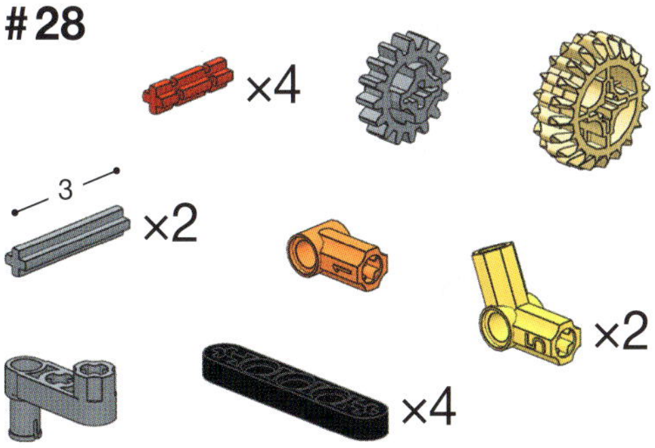

3

×2

×2

×4

SPEED POWER
UP DOWN

5:4 (20:16)

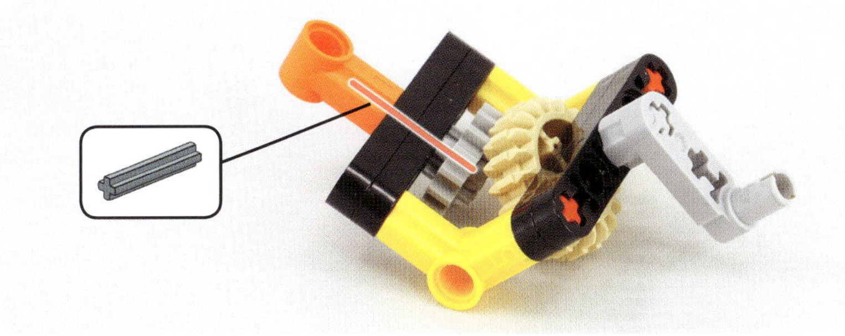

29

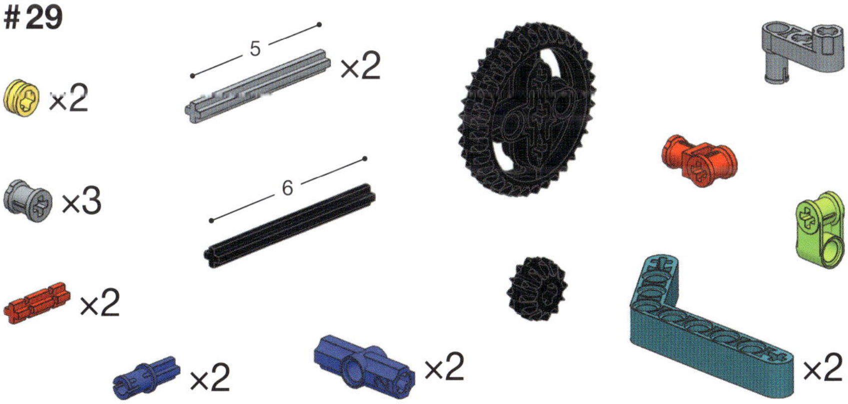

×2

5
×2

×3

6

×2

×2

×2

×2

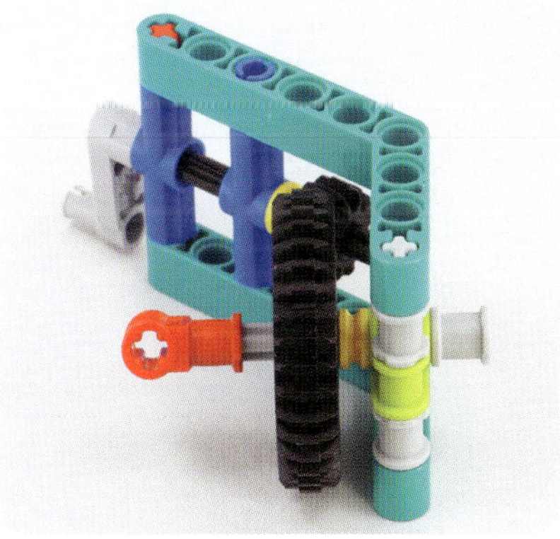

SLOW POWER
DOWN UP

1:3 (12:36)

Using worm gears

#30

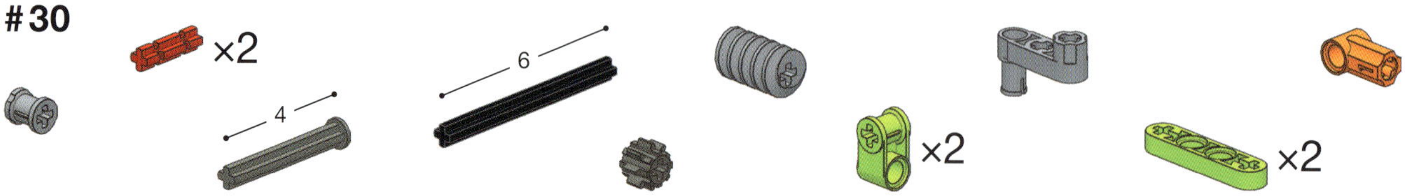

×2

4

6

×2

×2

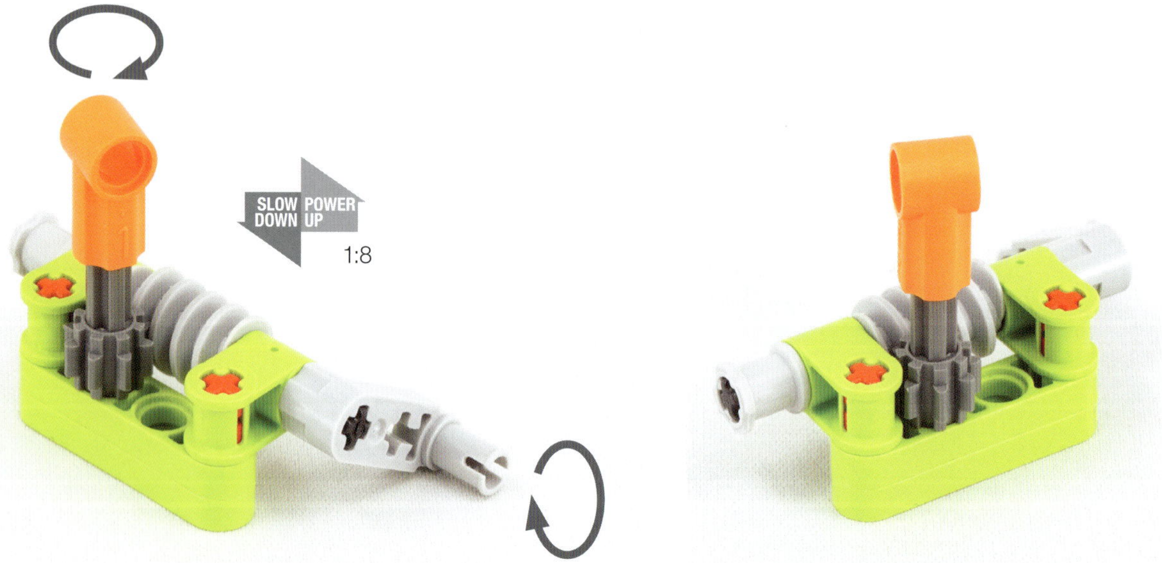

SLOW
DOWN
POWER
UP

1:8

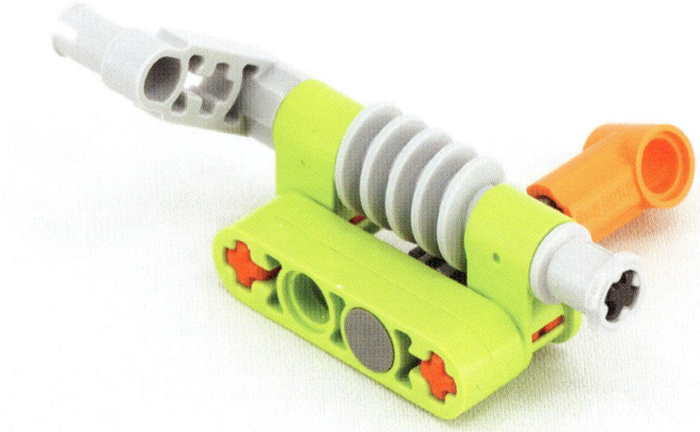

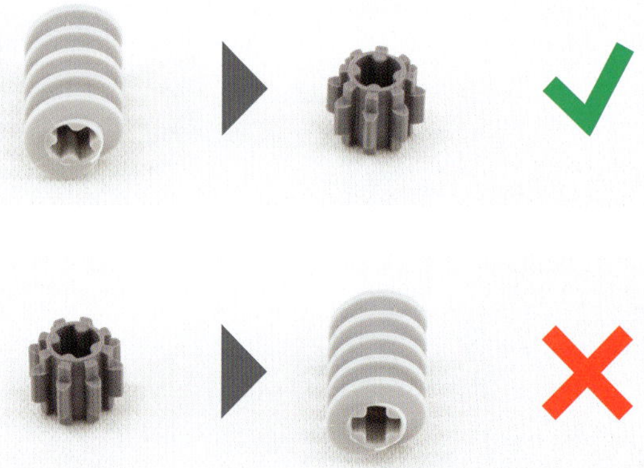

You can rotate an 8-tooth gear with a worm gear, but you can't rotate a worm gear with an 8-tooth gear.

24

×2

×2

4

6

×2

×2

SLOW POWER
DOWN UP
1:24

#32

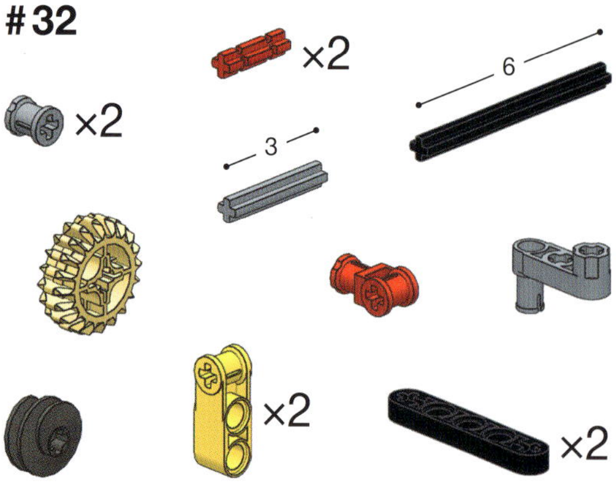

×2

×2

3

6

×2

×2

×2

SLOW POWER
DOWN UP
1:20

#33

4

5

×2

SLOW POWER
DOWN UP
1:36

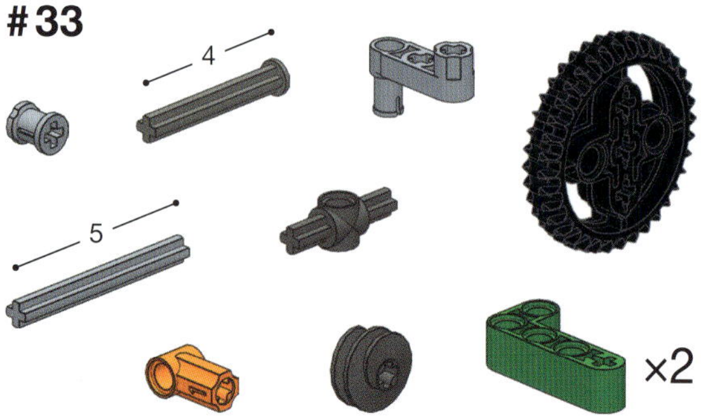

#34

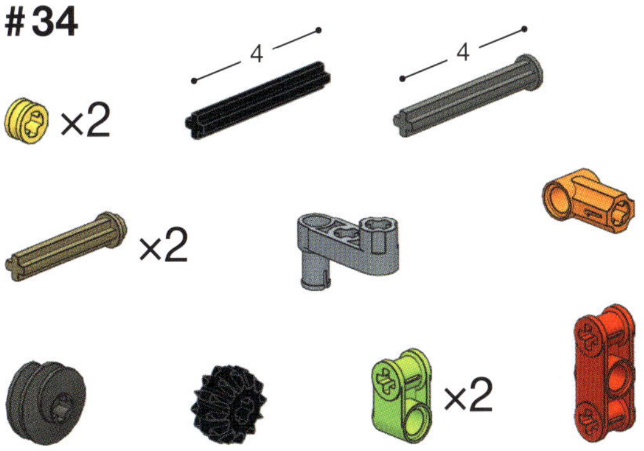

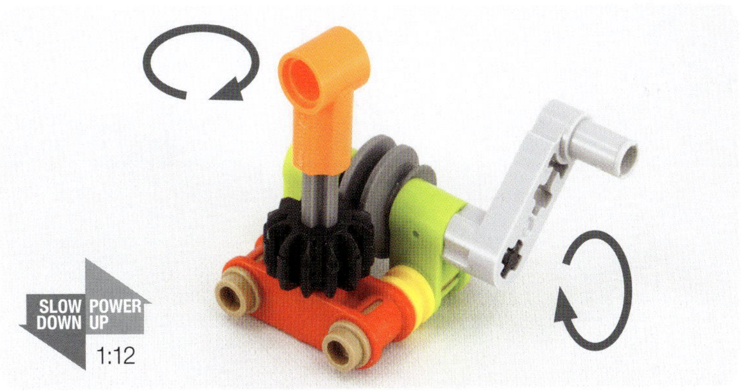

×2

4

4

×2

×2

×2

SLOW POWER
DOWN UP
1:12

#35

×2

4

6

×2

3
×2

×2

×2

×2

SLOW POWER
DOWN UP
1:16

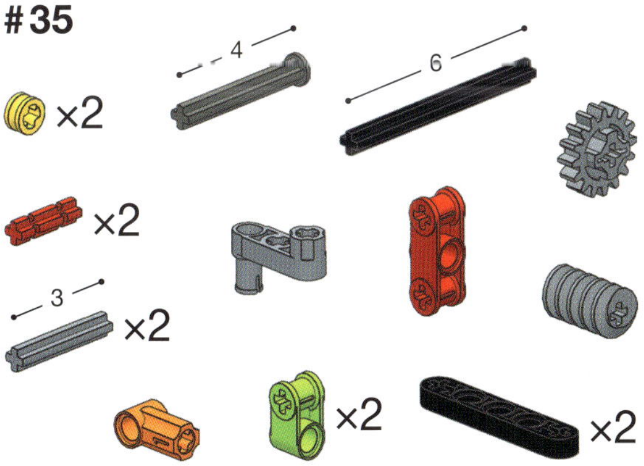

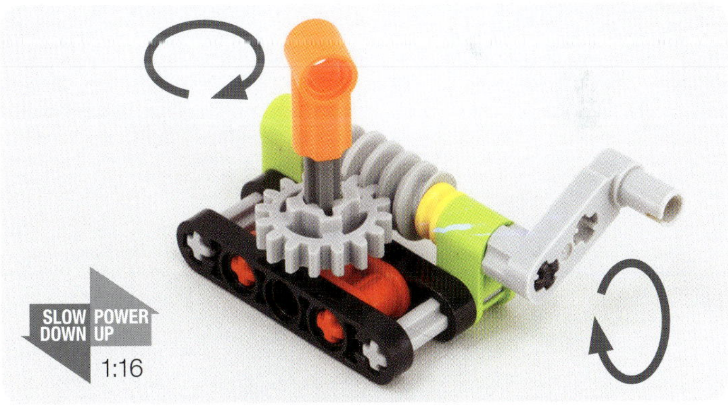

#36

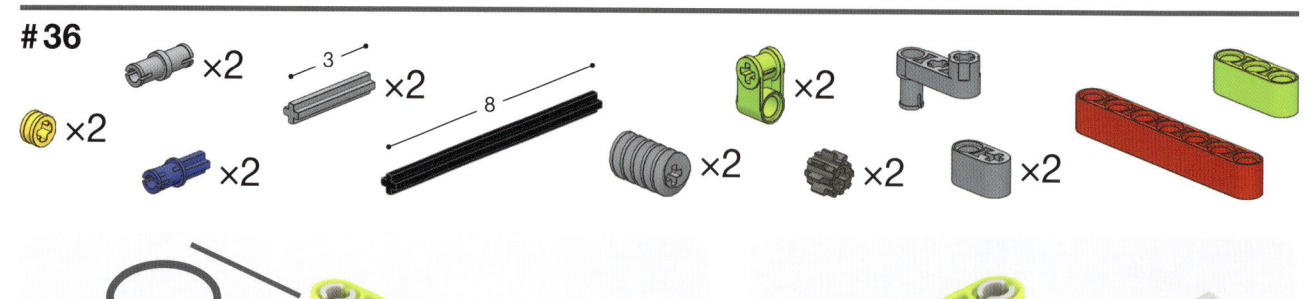

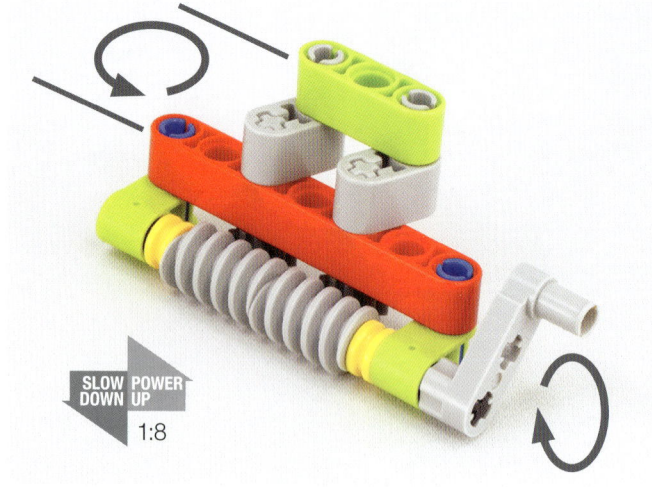

SLOW POWER
DOWN UP
1:8

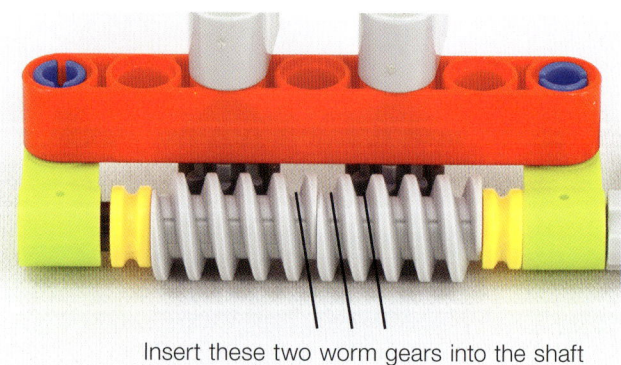

Insert these two worm gears into the shaft
so that their teeth connect smoothly.

#37

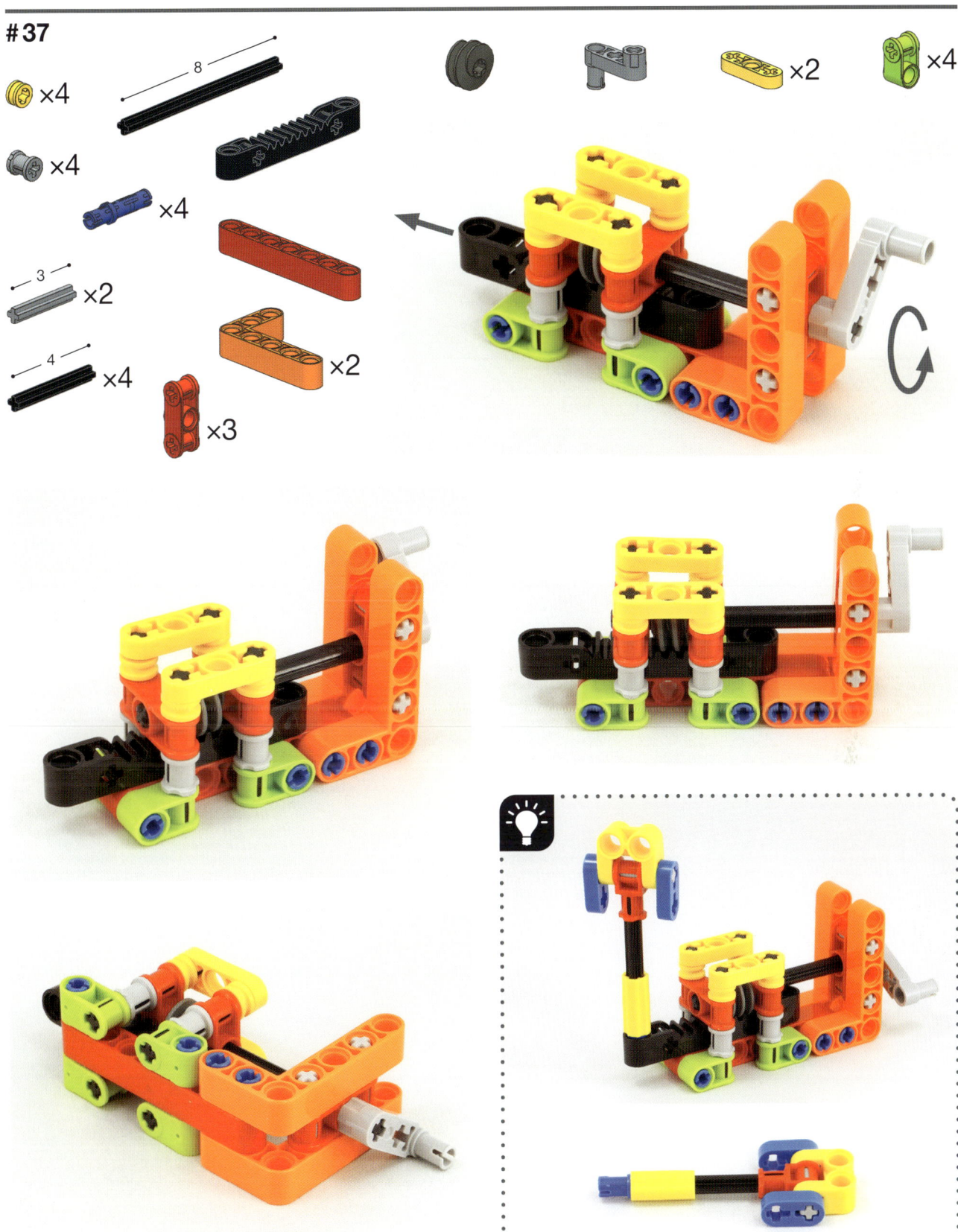

Using turntables

#38

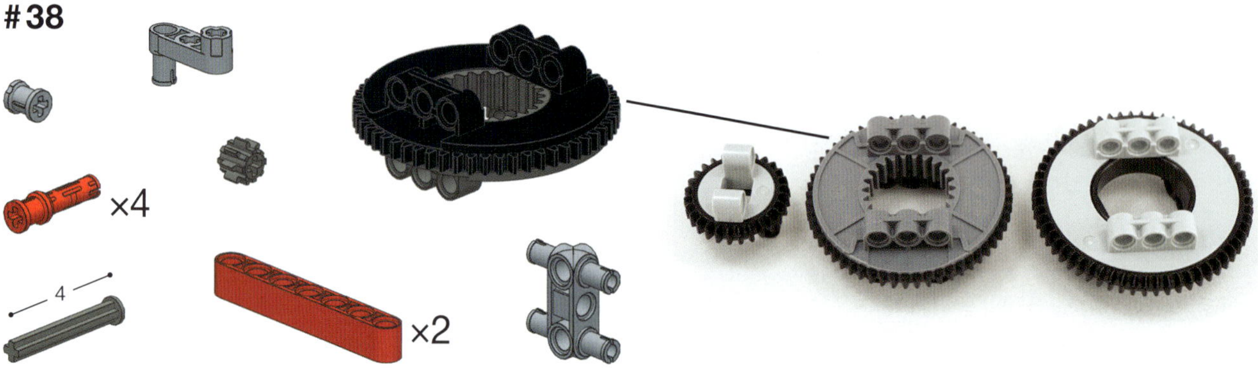

SLOW POWER
DOWN UP
1:7 (8:56)

×2

×8

3

×2

×2

1:5 (12:60)

SLOW POWER
DOWN UP

#40

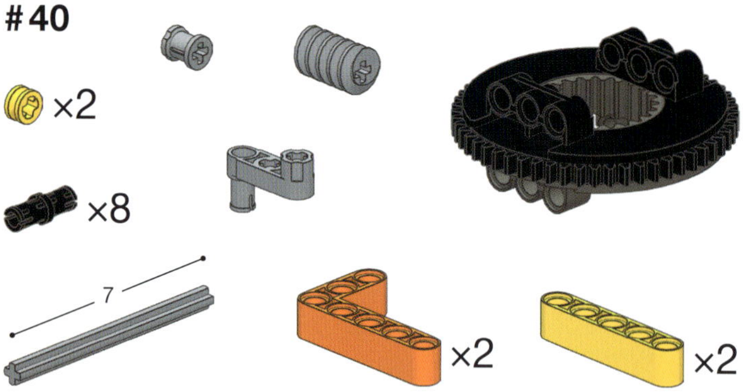

×2

×8

7

×2 ×2

SLOW POWER
DOWN UP
1:56

SLOW POWER
DOWN UP
1:60

×6

6

×2

×2

×2

×2

#42

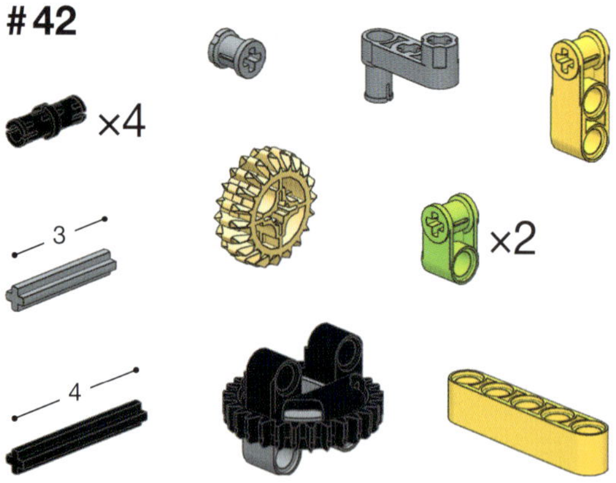

×4

3

4

×2

SLOW DOWN **POWER UP**

5:7 (20:28)

#43

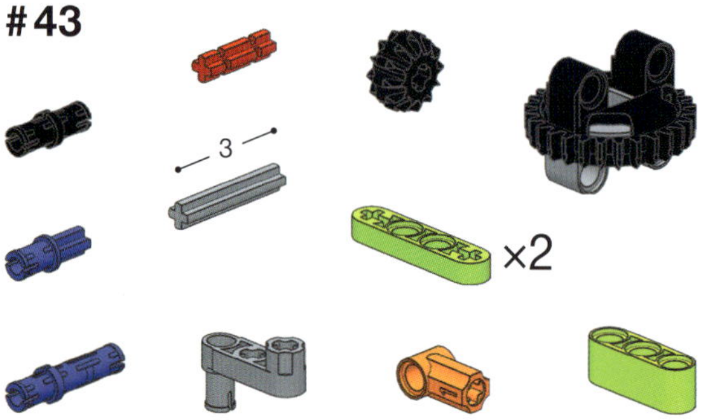

3

×2

SLOW DOWN **POWER UP**

3:7 (12:28)

#44

4

5

×2

×2 ×2

SLOW POWER
DOWN UP

1:3 (8:24)

Changing the angle of the axle

#45

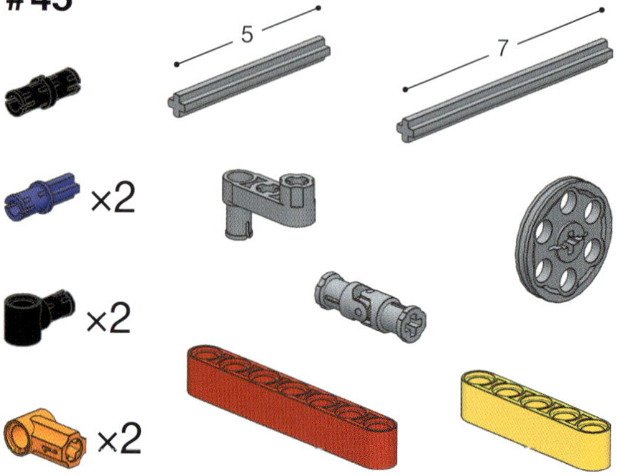

5

7

×2

×2

×2

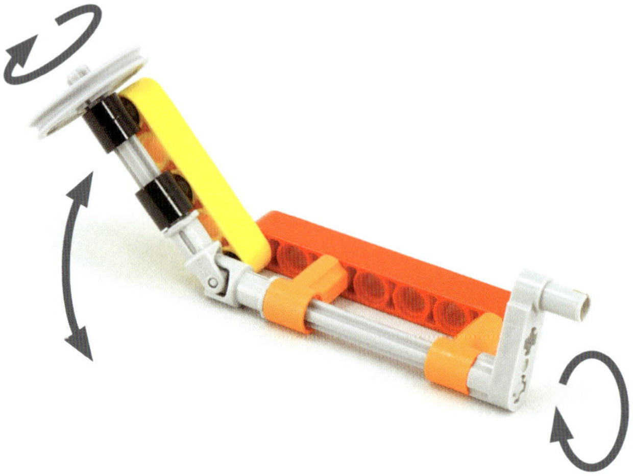

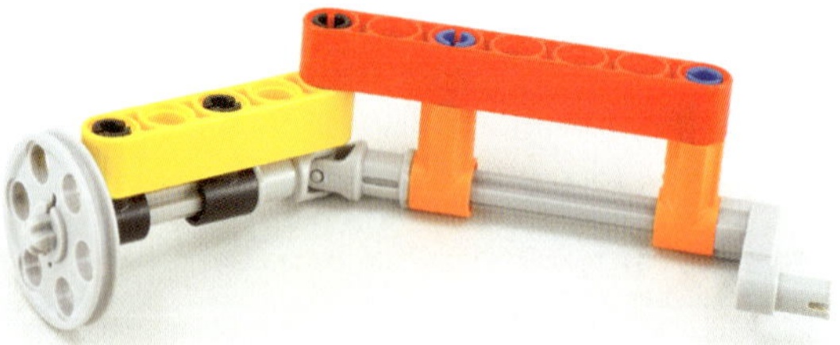

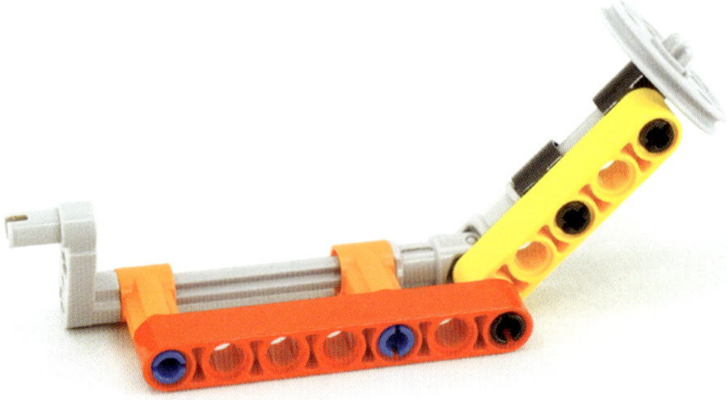

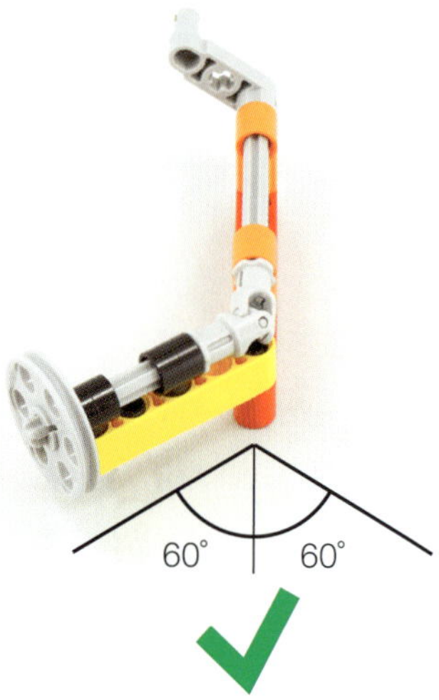

60° 60°

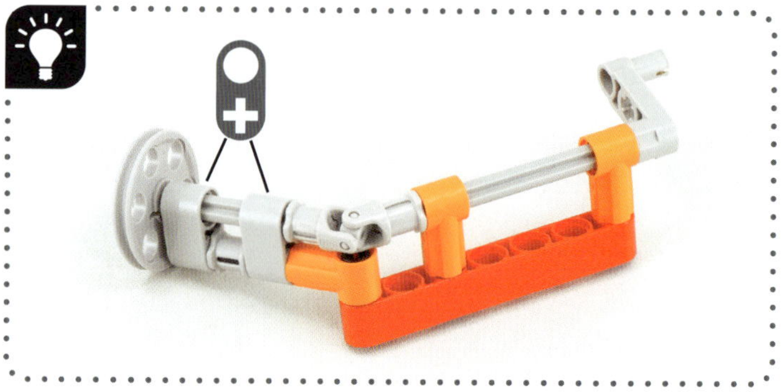

✔

#46

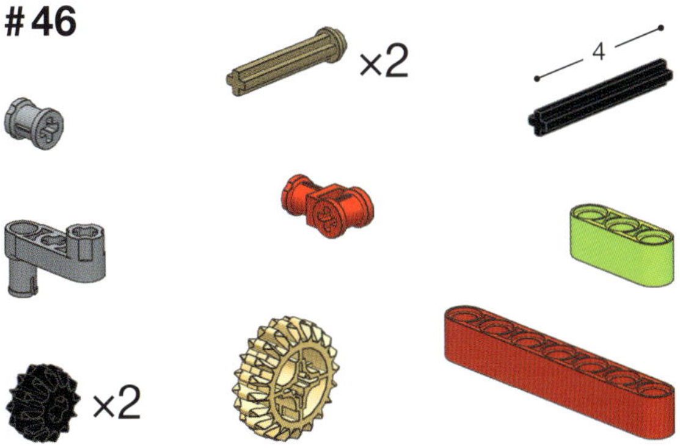
×2
4
×2

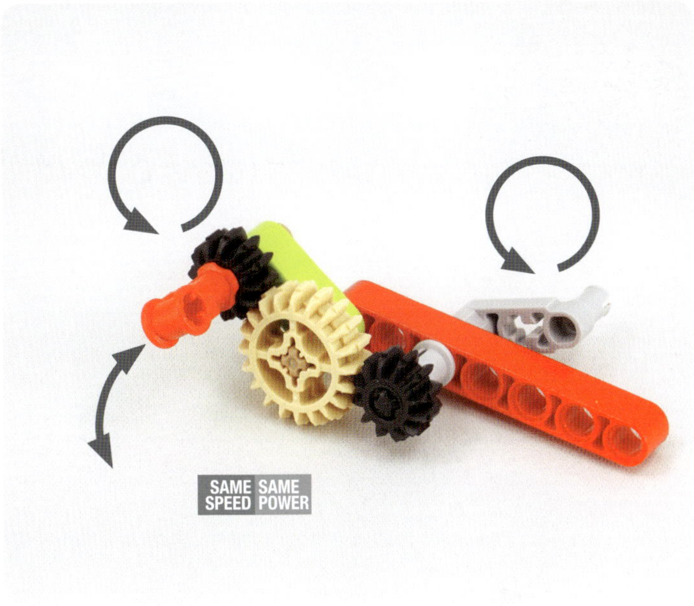
SAME SPEED
SAME POWER

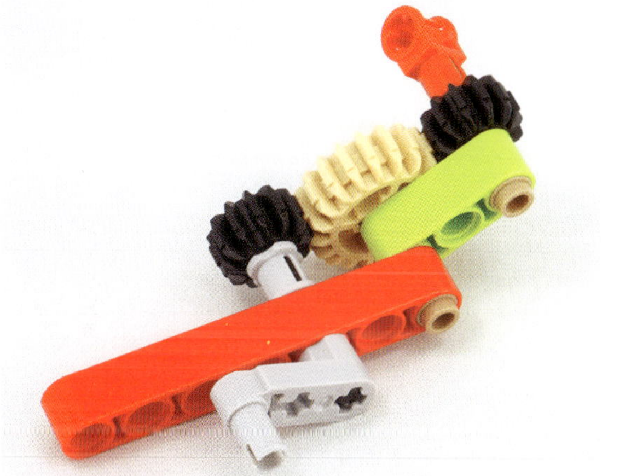

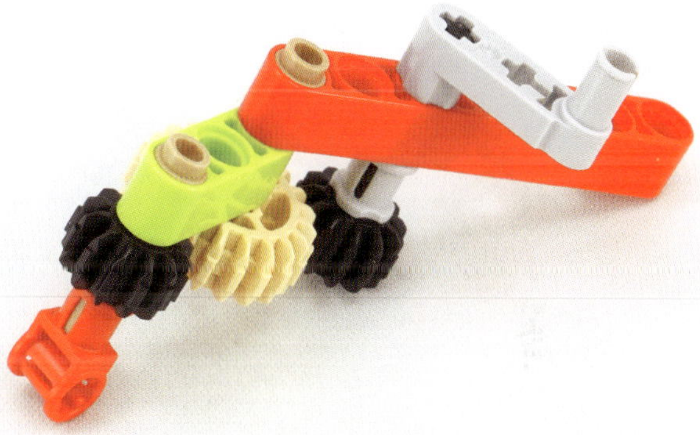

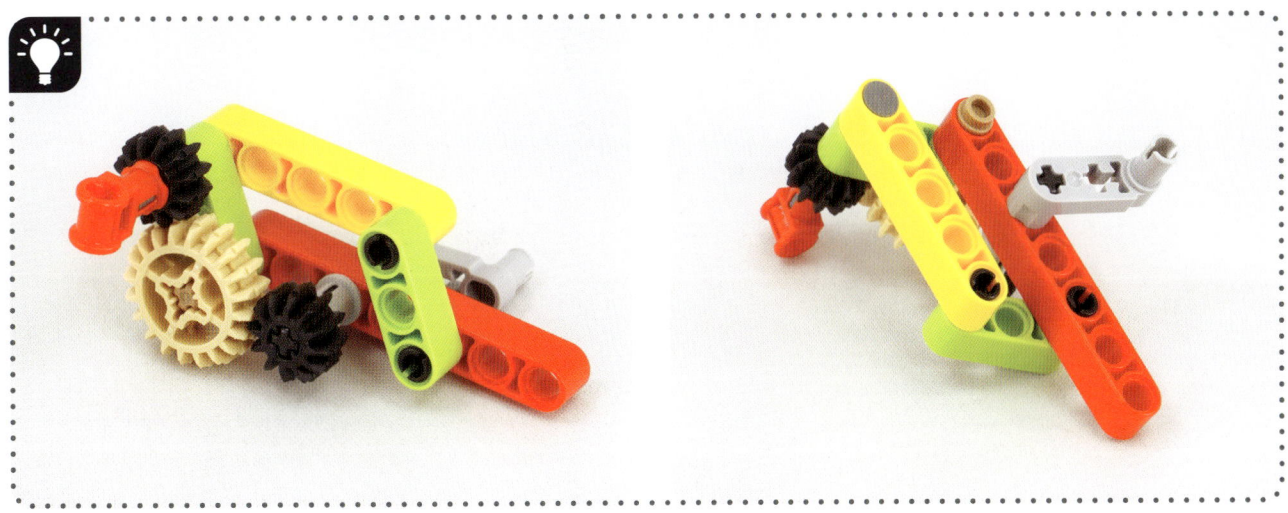

#47

×2

×2 ×4

×4

×2 ×2

5

6

SAME SPEED SAME POWER

×3

×2

×2

×3

5

×6

7

×2

SAME SAME
SPEED POWER

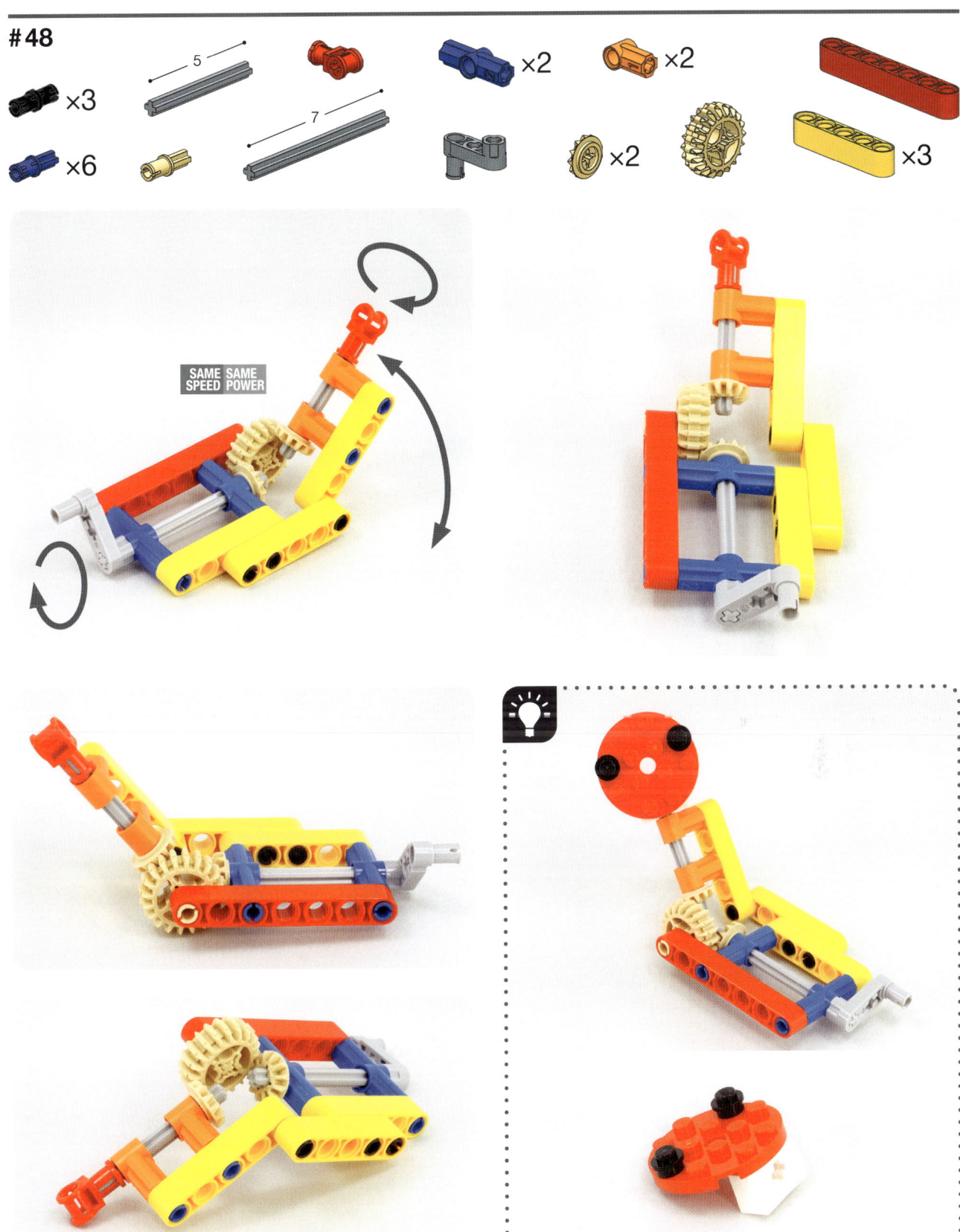

#49

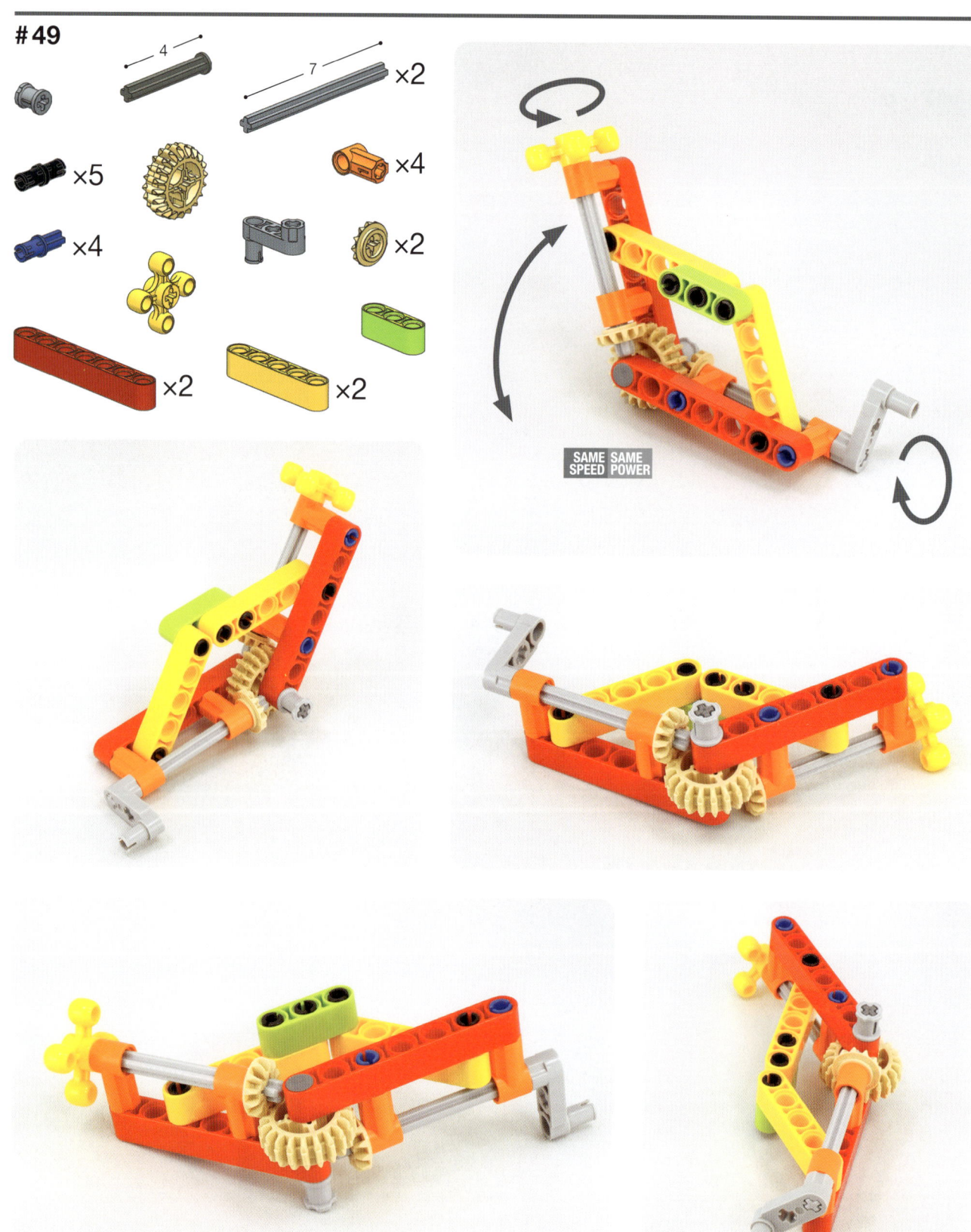

×2 · 4

×2 · 7

×5

×4

×4

×2

×2

×2

SAME SPEED SAME POWER

×2

×2

×4

×2

×6

×2

×2

SAME SPEED SAME POWER

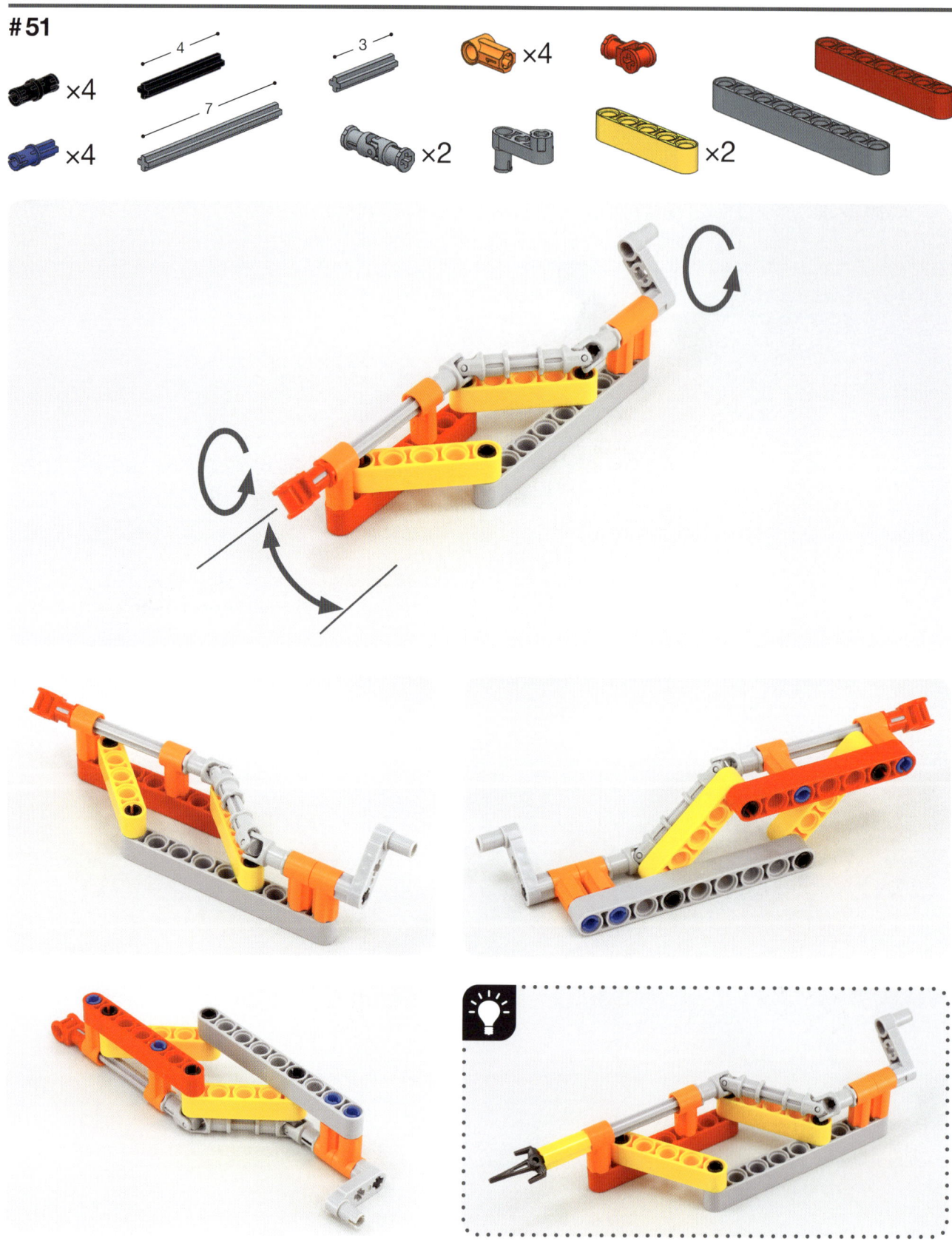

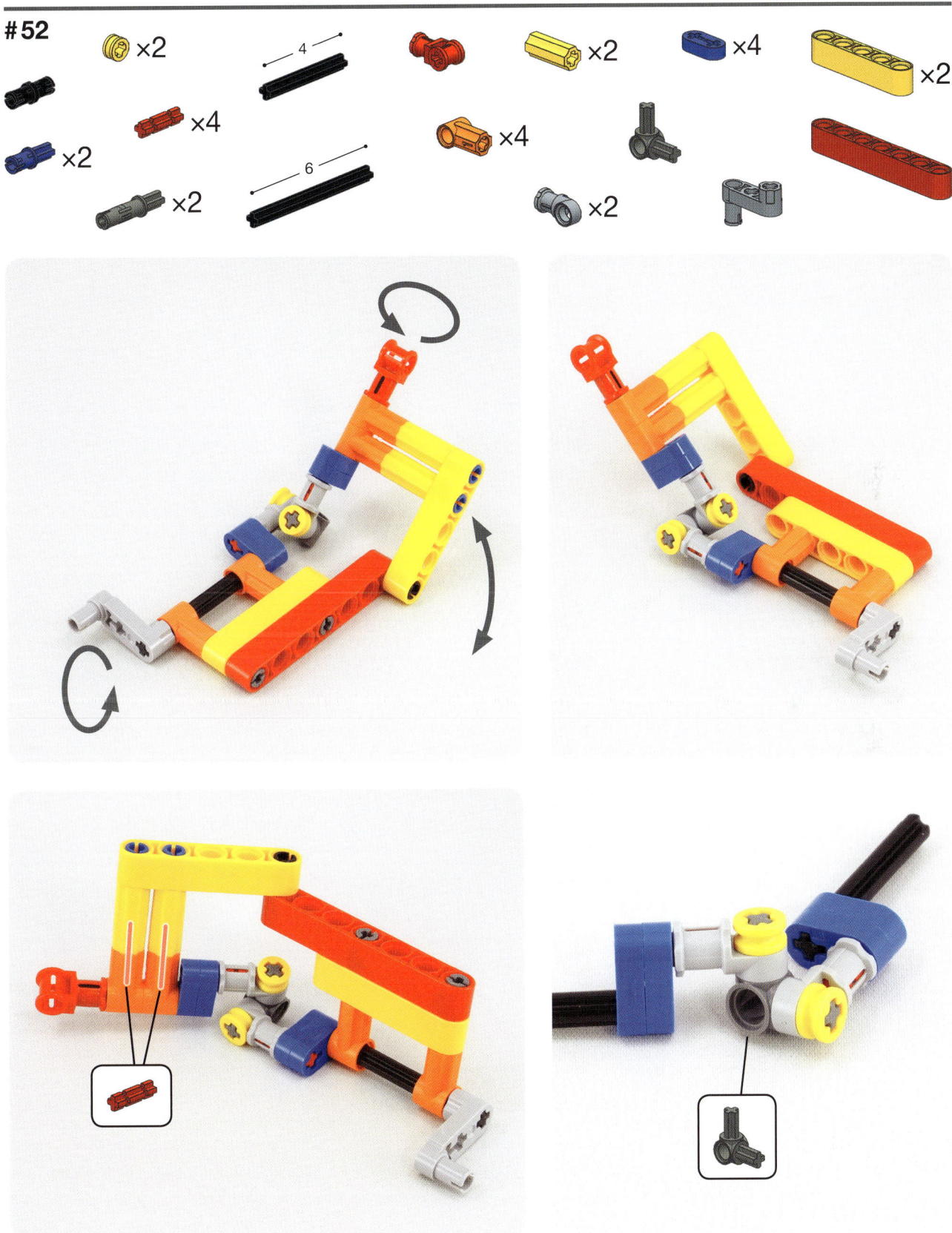

Transmitting rotation with rubber bands

#53

 ×2  ×2  ×2

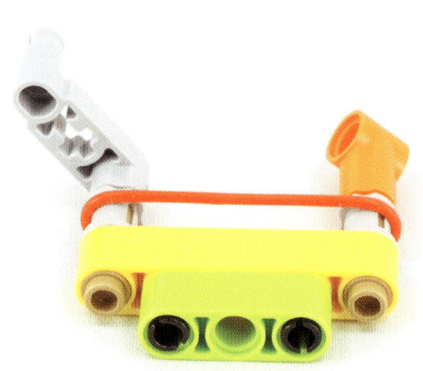

#54

×2   3

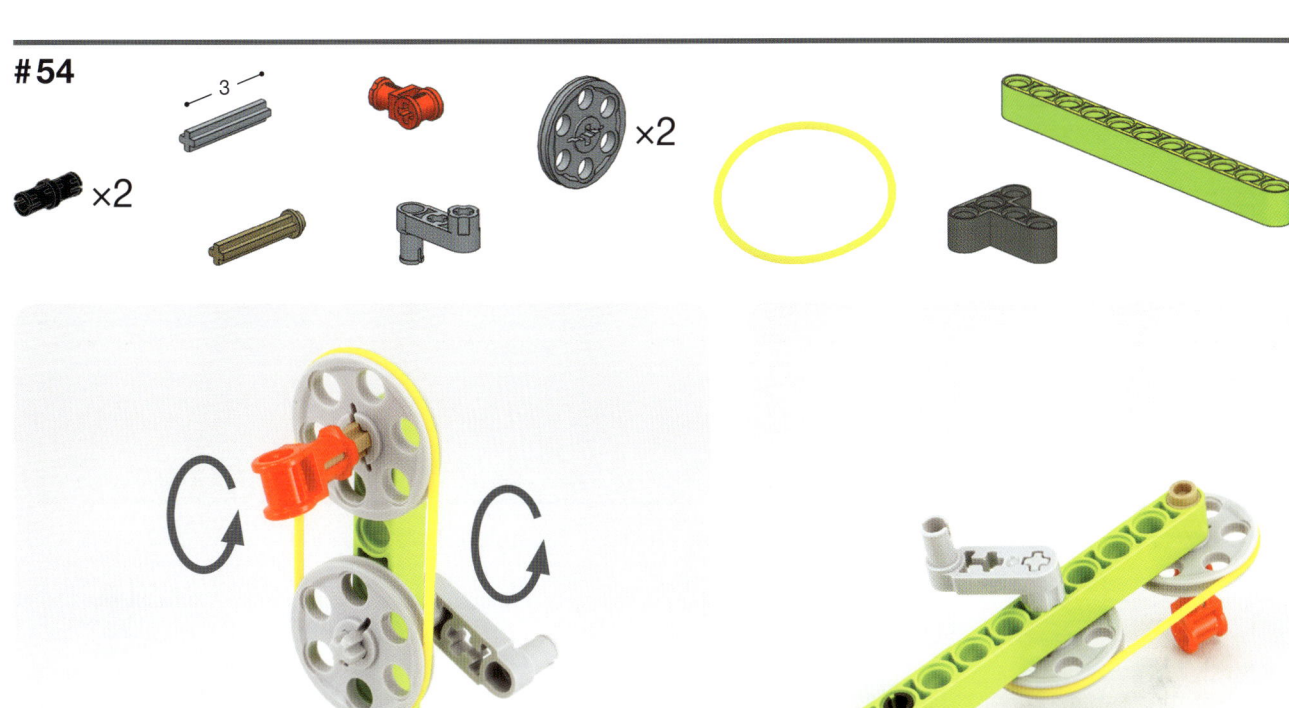

SLOW
DOWN

SPEED
UP

#55

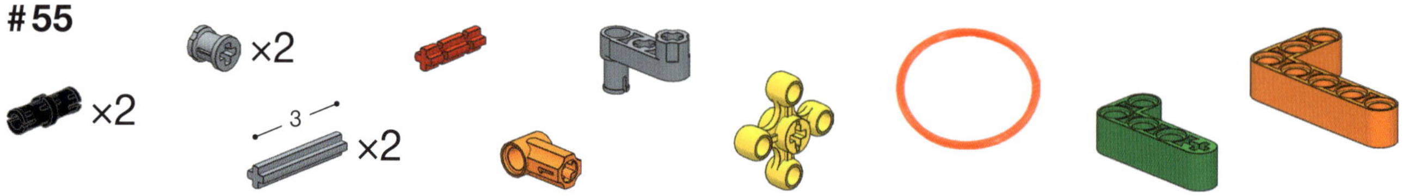

×2 ×2 —3—· ×2

#56

  ×2

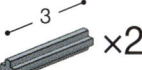 ×2

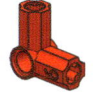

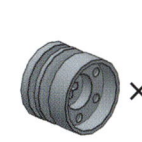 ×2

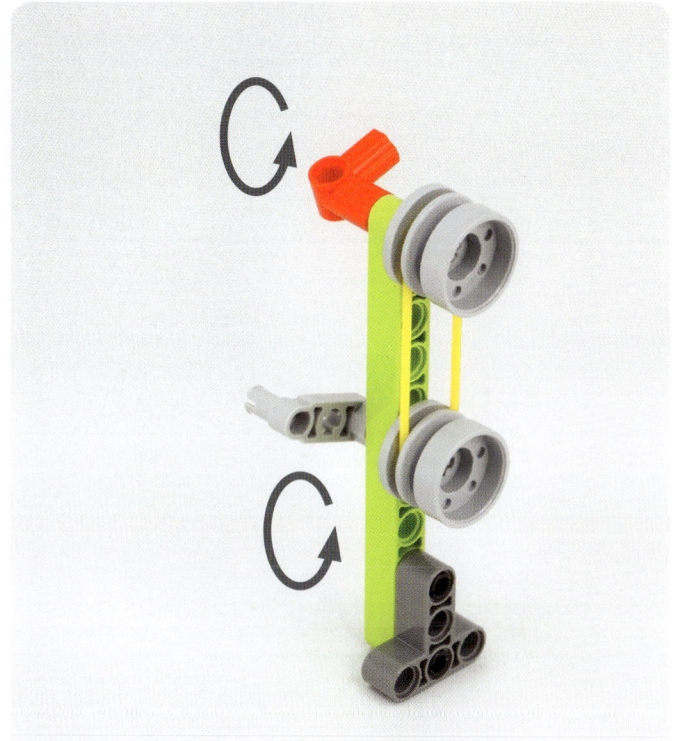

When you use two rubber bands, more power can be transmitted.

#57

 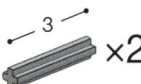 ×2  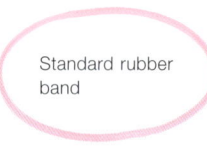 ×3

Standard rubber band

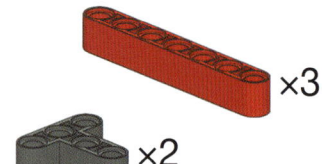 ×3

 ×8   ×2

×2

#58

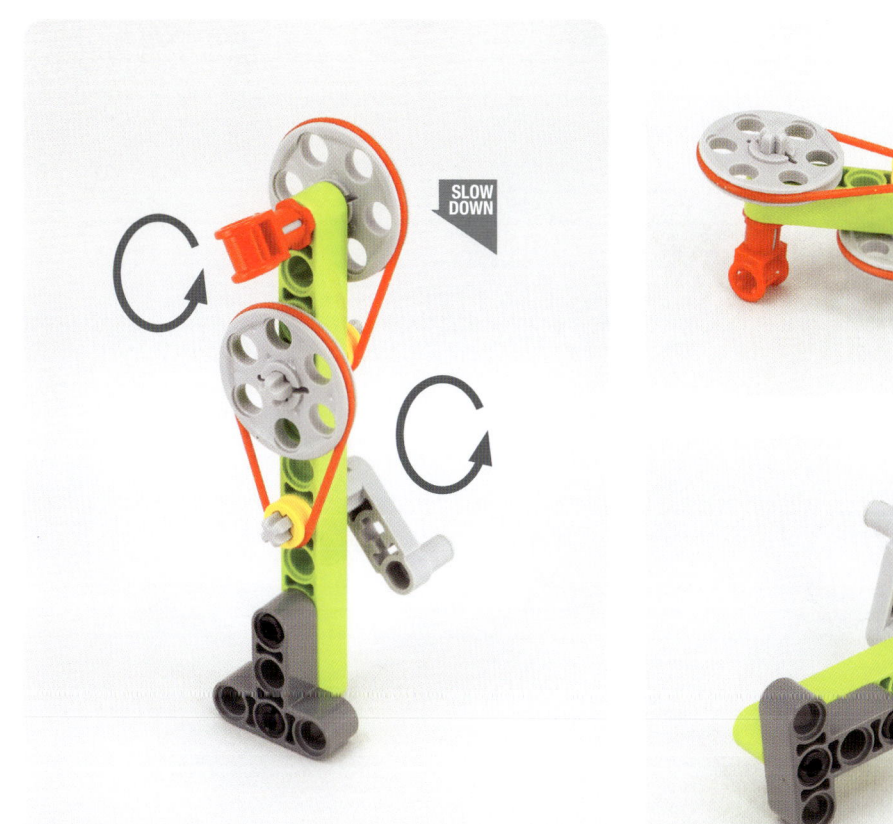

SLOW
DOWN

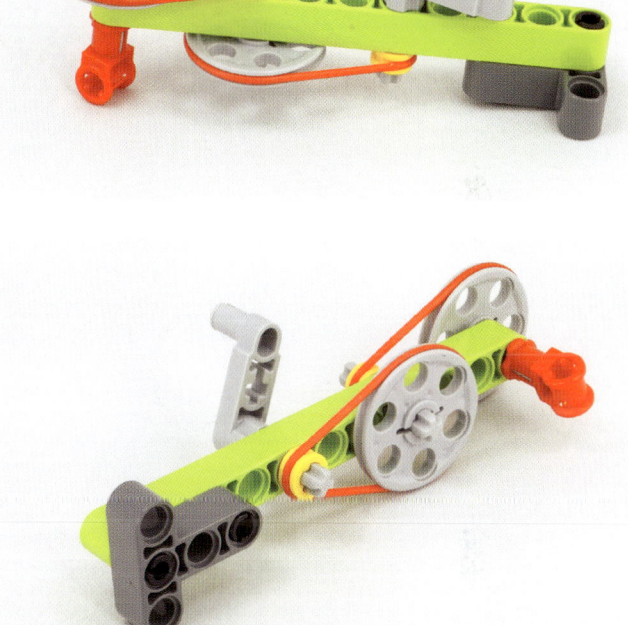

Transmitting rotation with chains and treads

#59

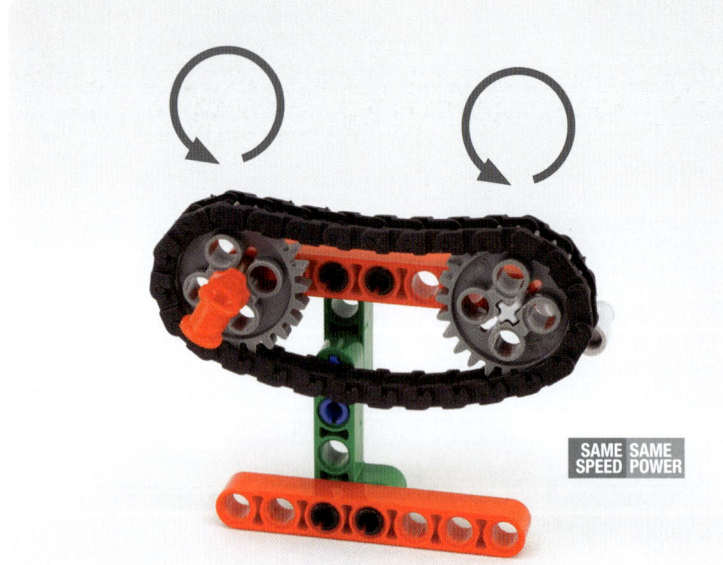

SAME SPEED | SAME POWER

#60

×2 ×2 ×27 4 4

SLOW DOWN POWER UP

2:5 (16:40)

SPEED UP POWER DOWN

5:2 (40:16)

SLOW DOWN POWER UP

2:3 (16:24)

×3

×4

4

6

×31

SPEED UP POWER DOWN

5:3 (40:24)

#62

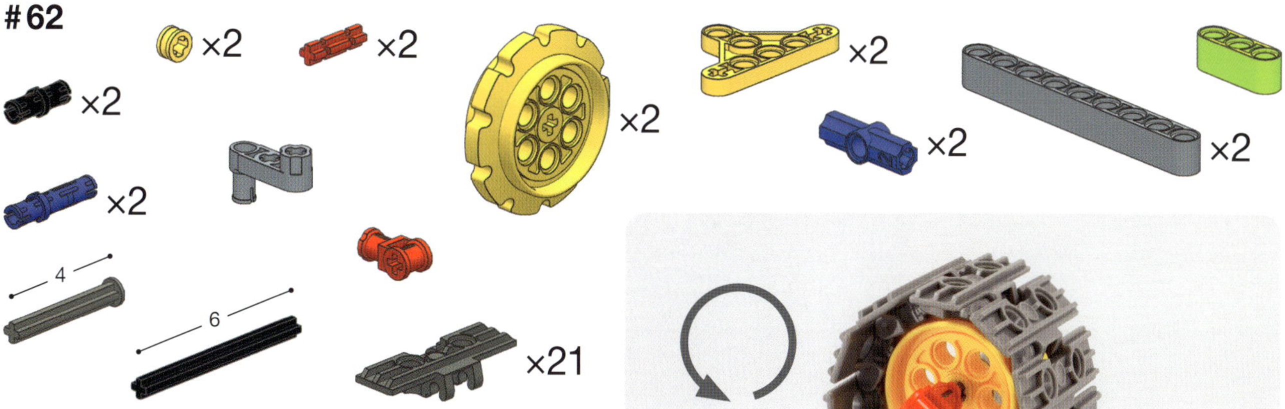

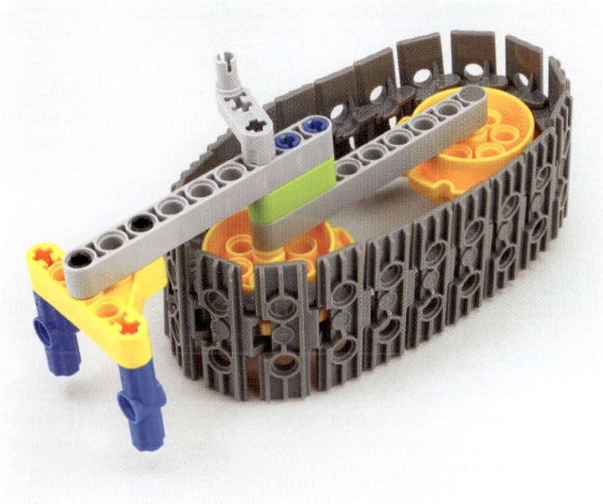

×2　×2

×2

×2

×2

×2

×2

×2

4

6

×21

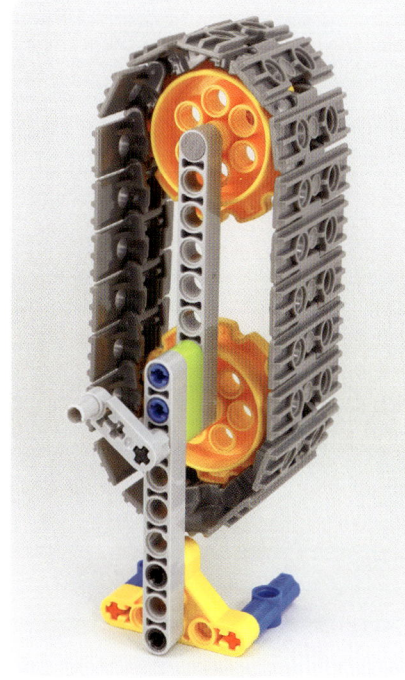

SAME SAME
SPEED POWER

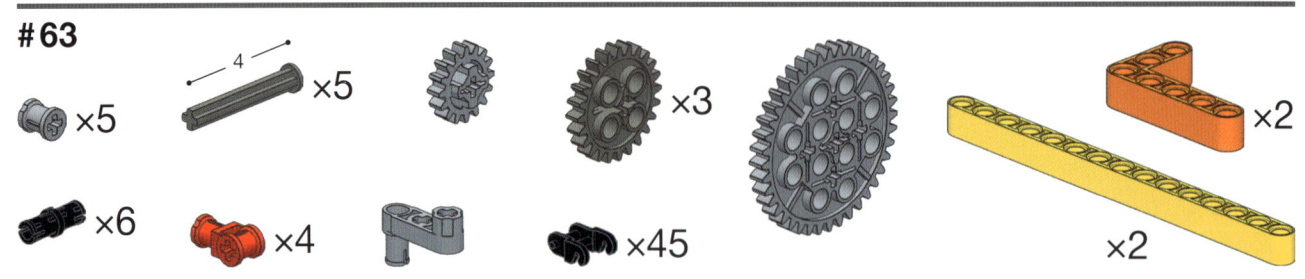

×5   ×5   ×3   ×2   ×2

×6   ×4   ×45

×2

—3—

—4—

—4—

×2

×35

×2

SAME SPEED SAME POWER

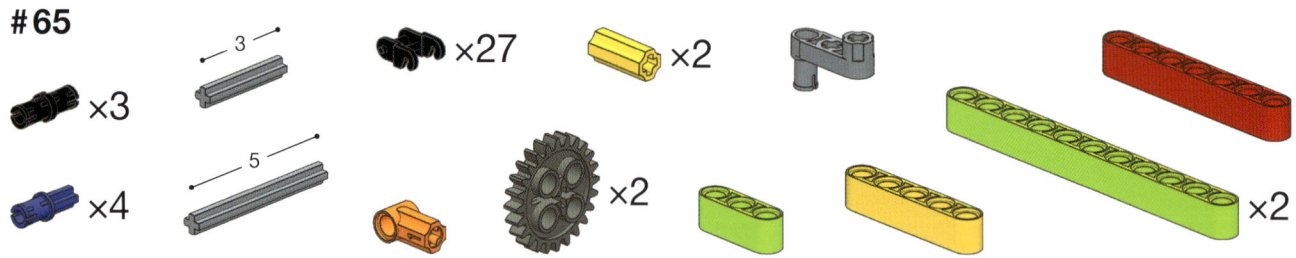

×3 ×3 ×27 ×2

×4 5 ×2 ×2

SAME SAME
SPEED POWER

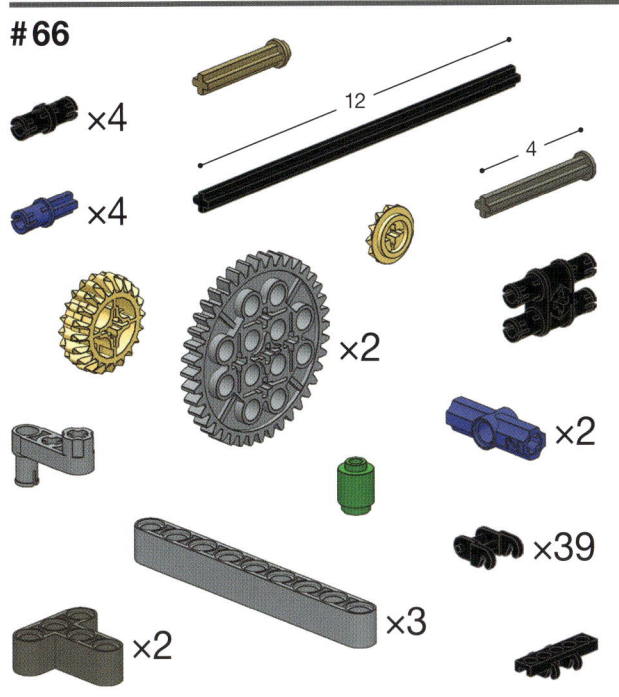

#66

×4

×4

12

4

×2

×2

×39

×3

×2

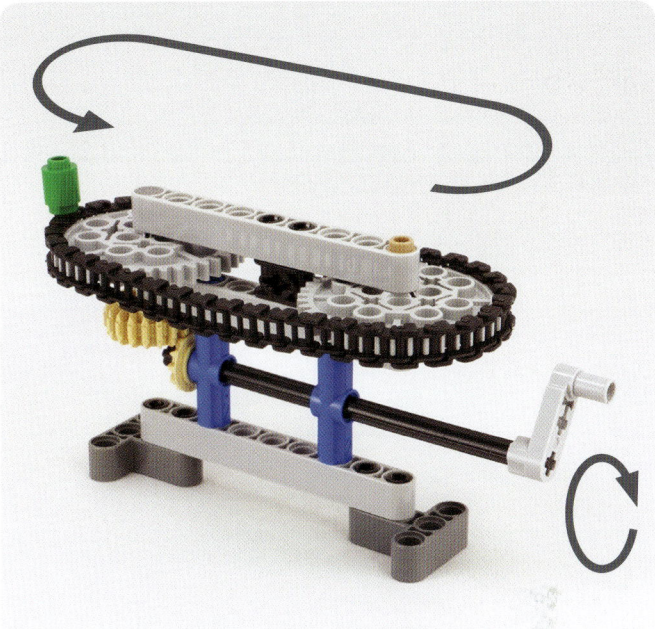

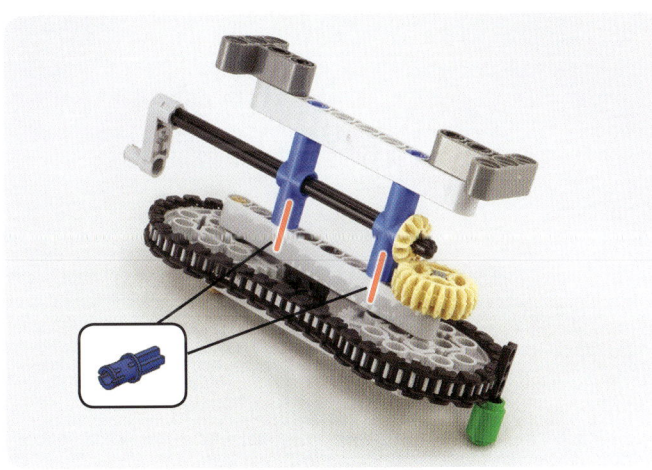

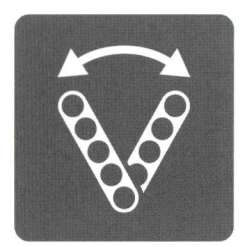

Oscillating mechanisms

#67

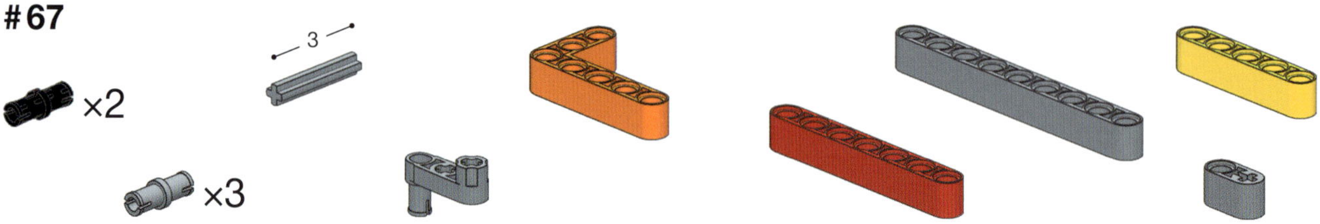

×2

×3

#68

×2 ×3 3

#71

#72

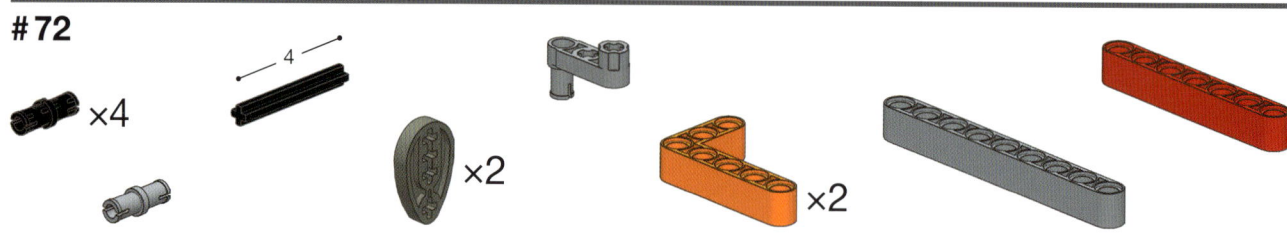

×4

×2

×2

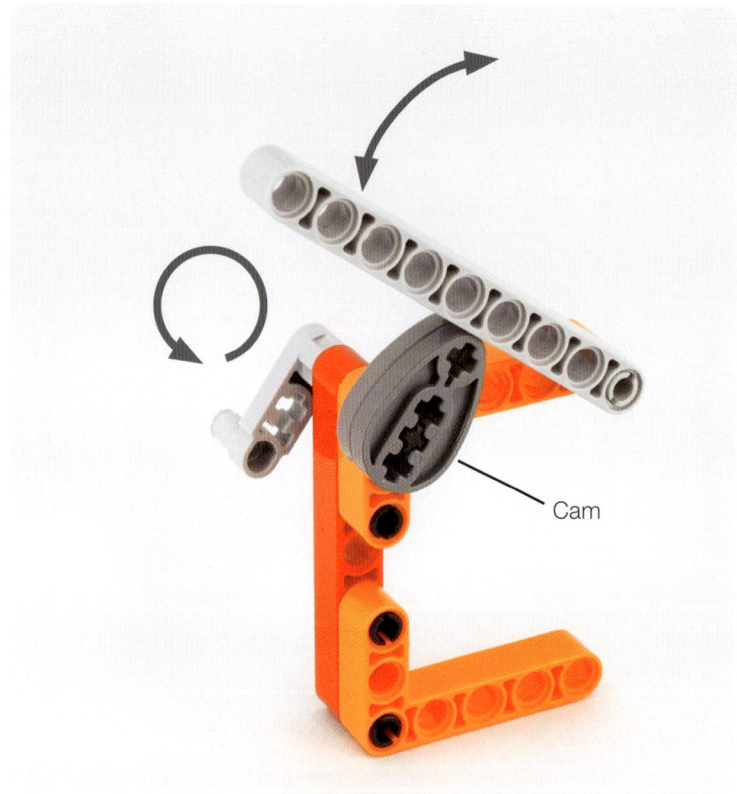

Cam

#73

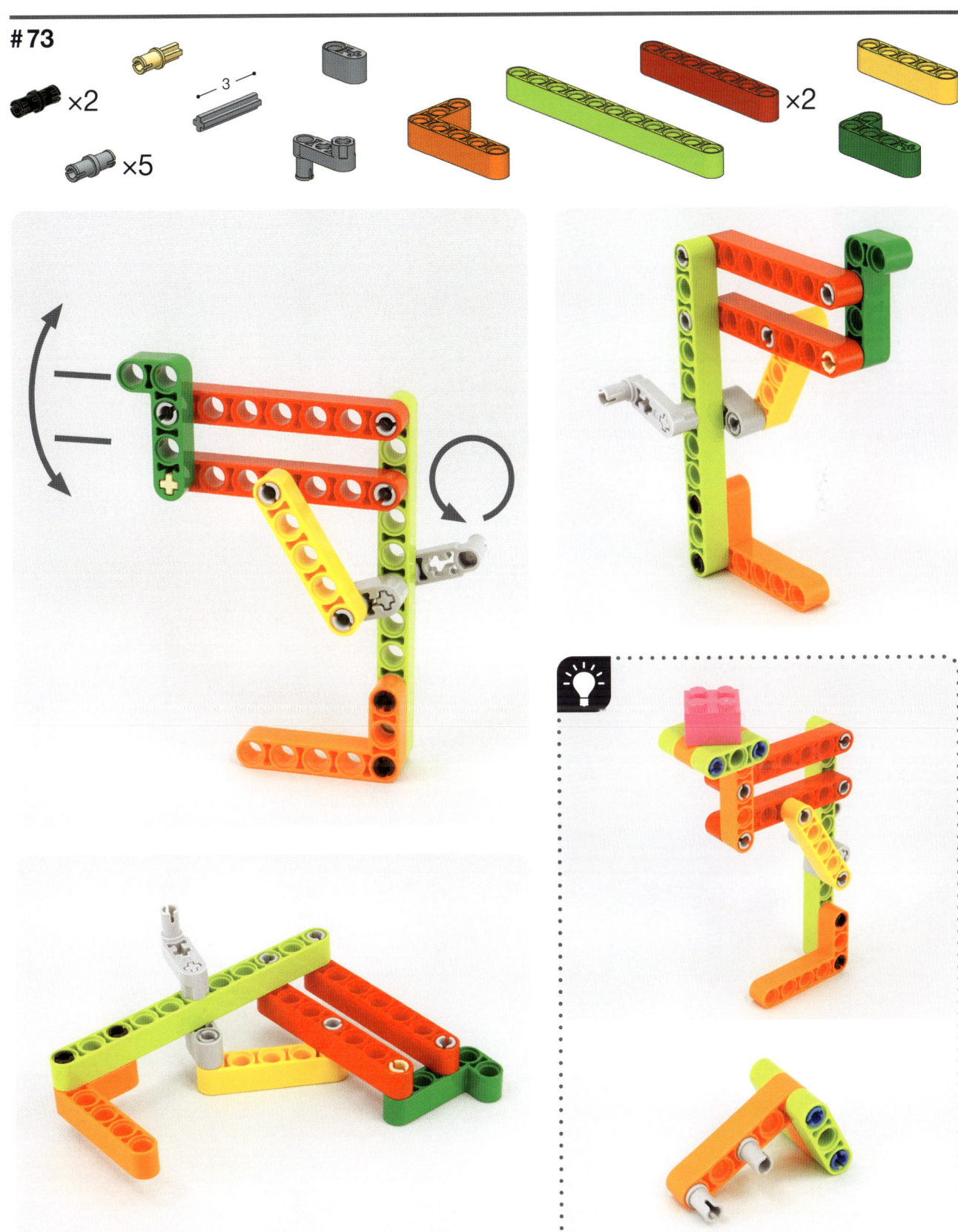

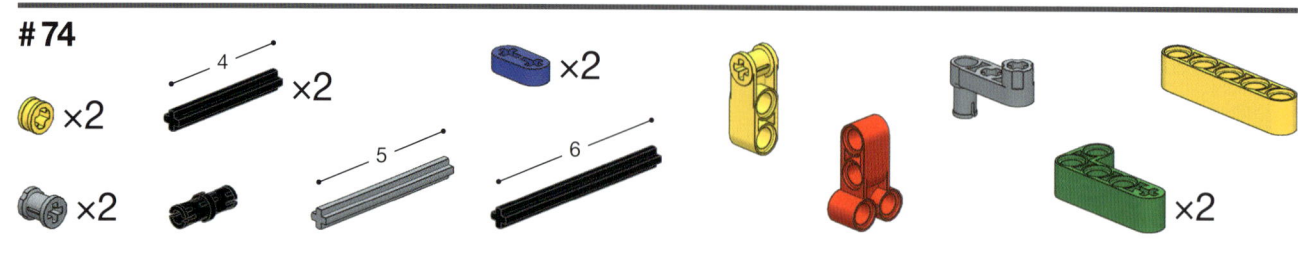

×2 ×2 ×2
×2 | 4 | 5 | 6

#75

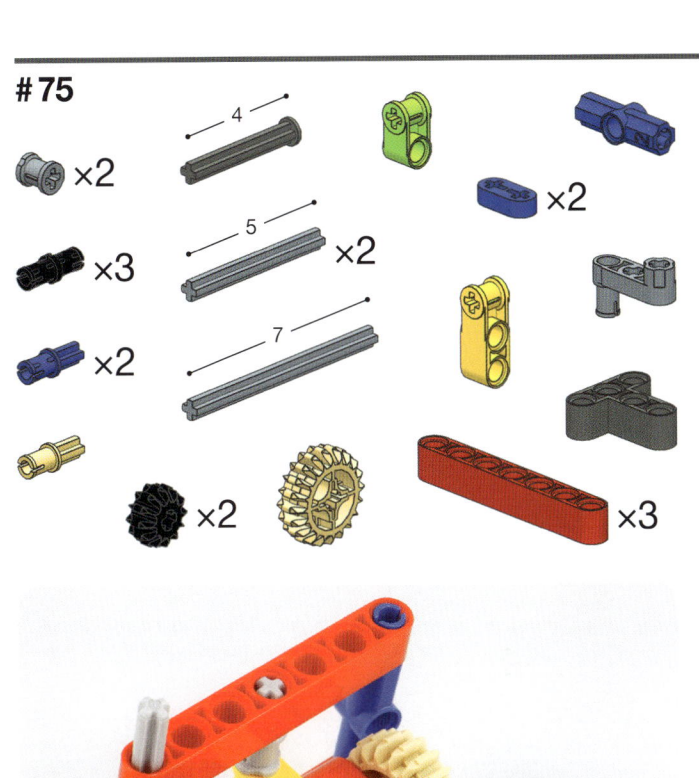

×2 ×3 ×2

4 5 7 ×2 ×2

×2 ×3

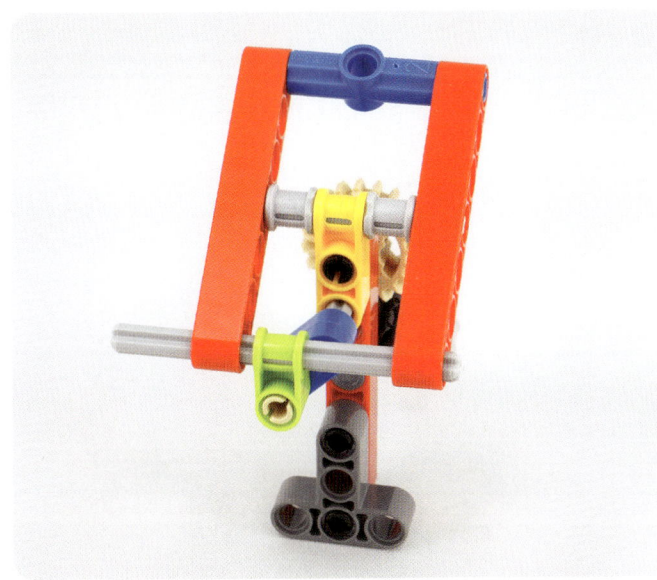

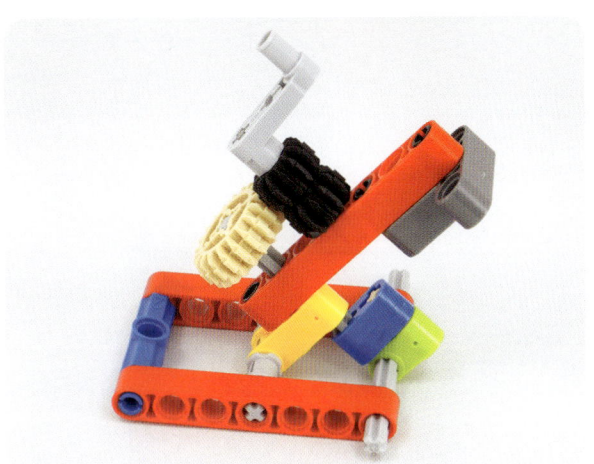

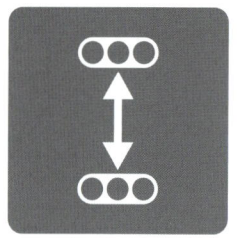

Reciprocating mechanisms

#76

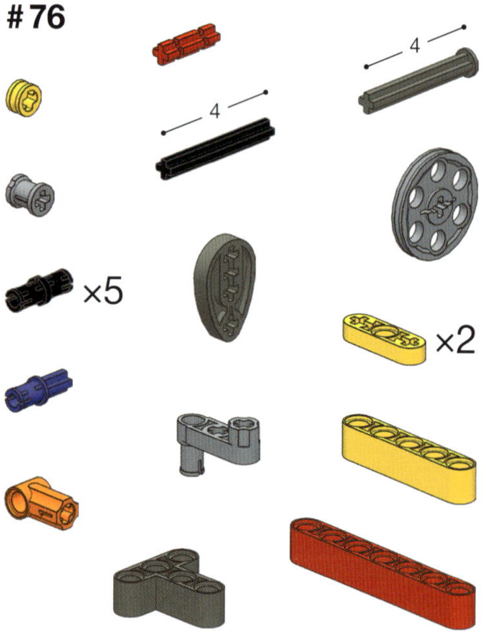

×5

×2

Cam

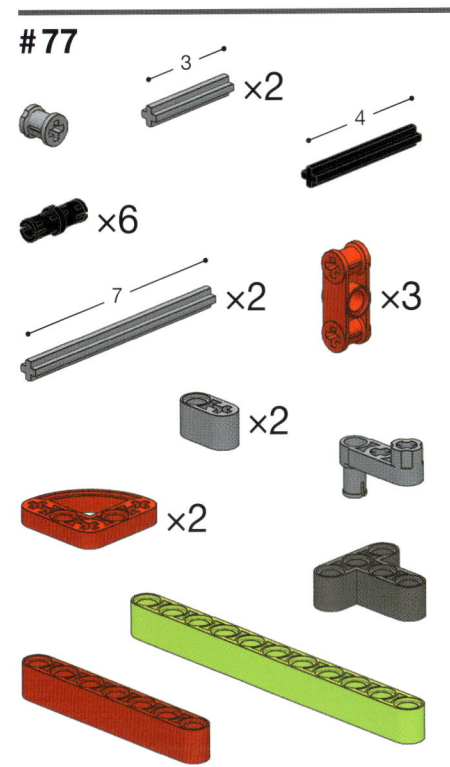

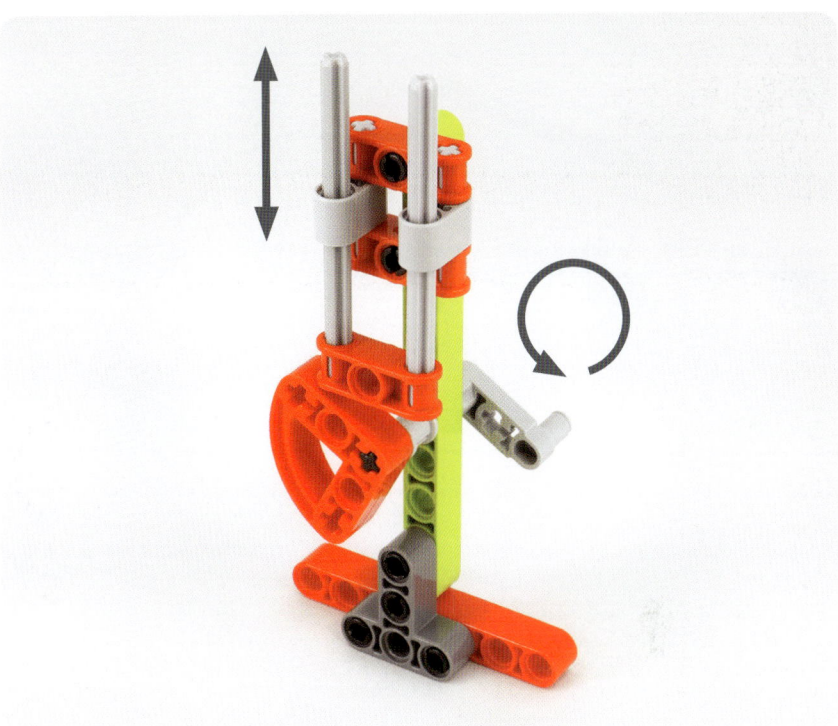

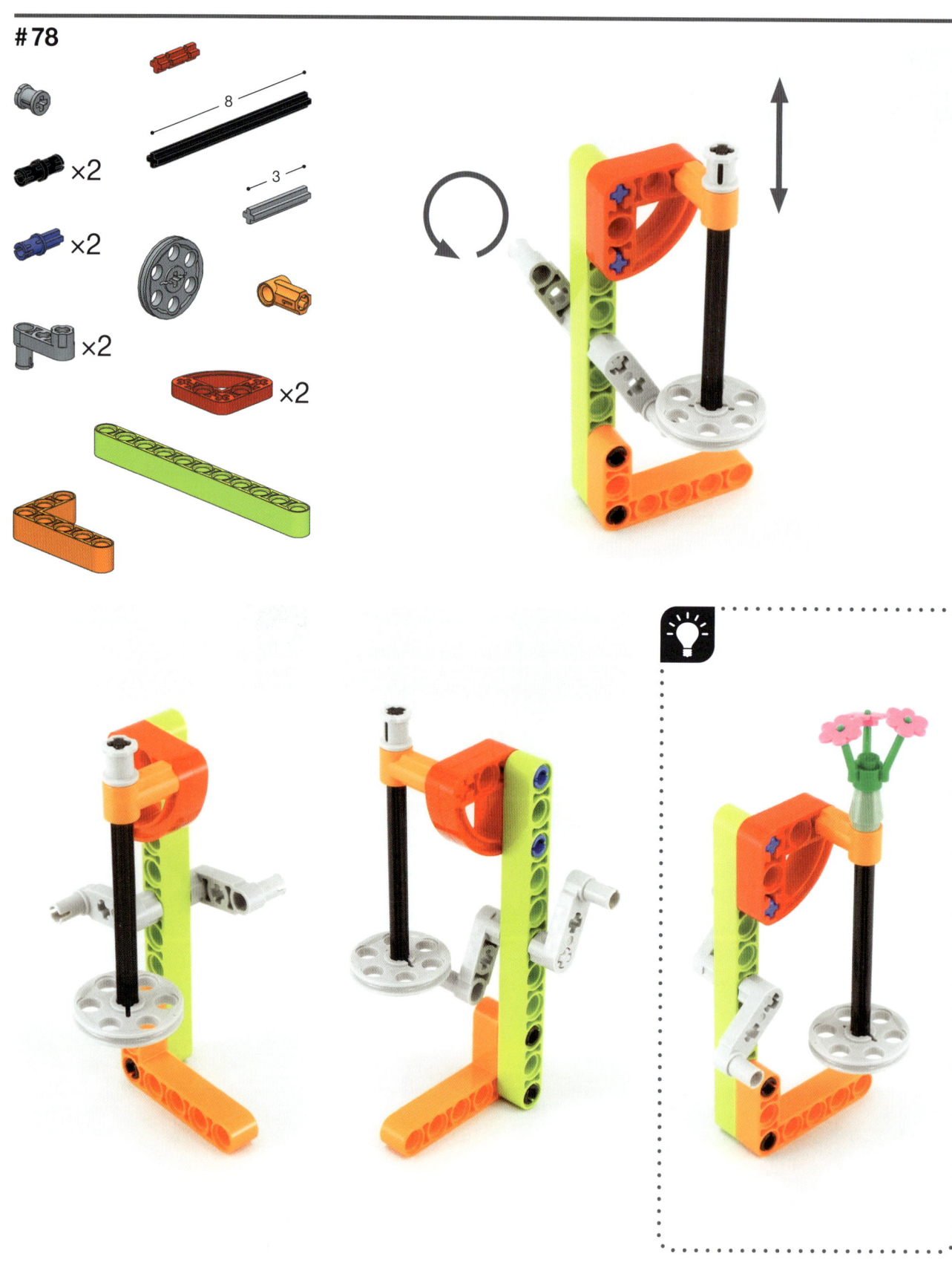

#79

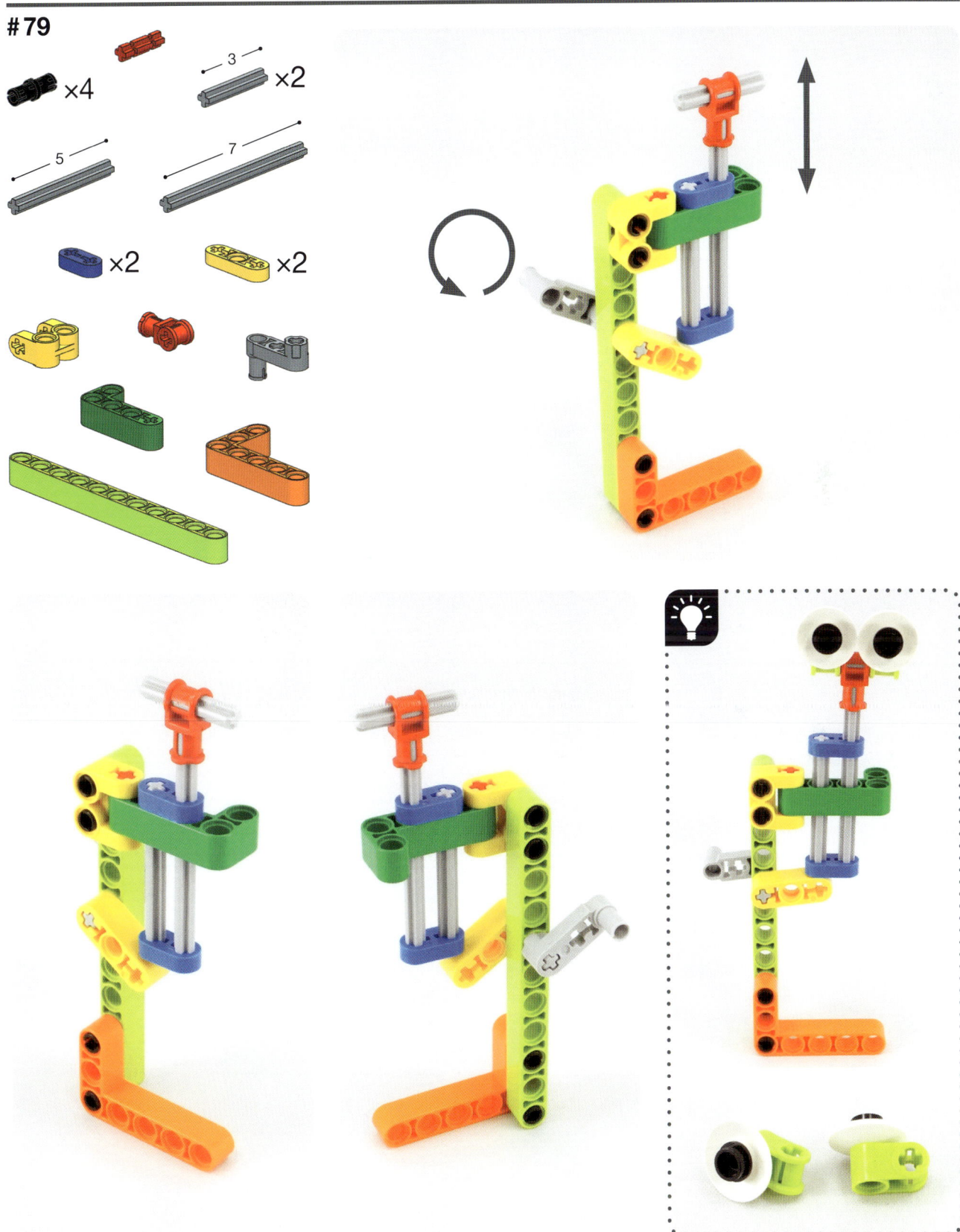

#80

#81

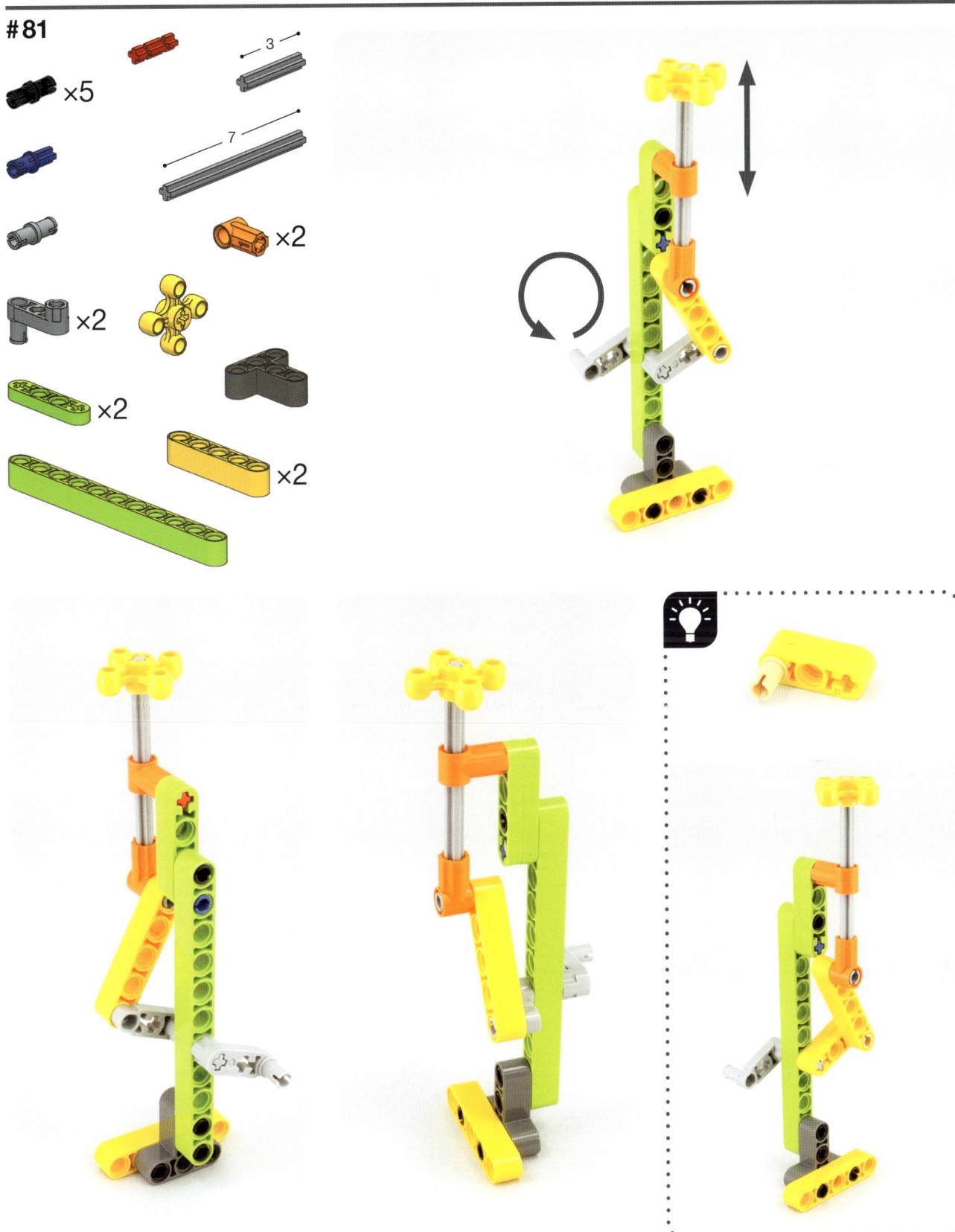

#82

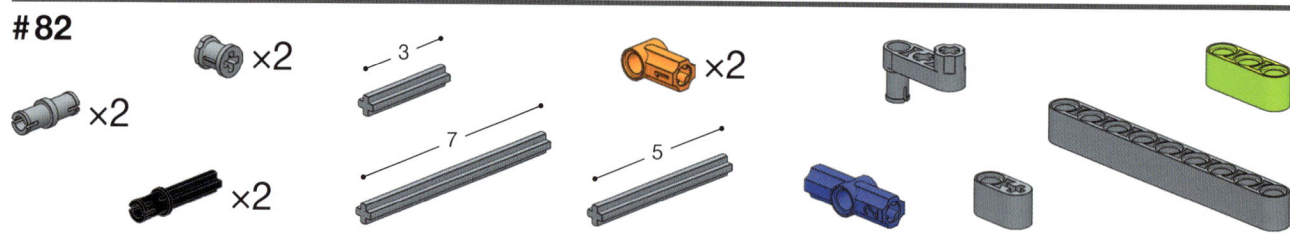

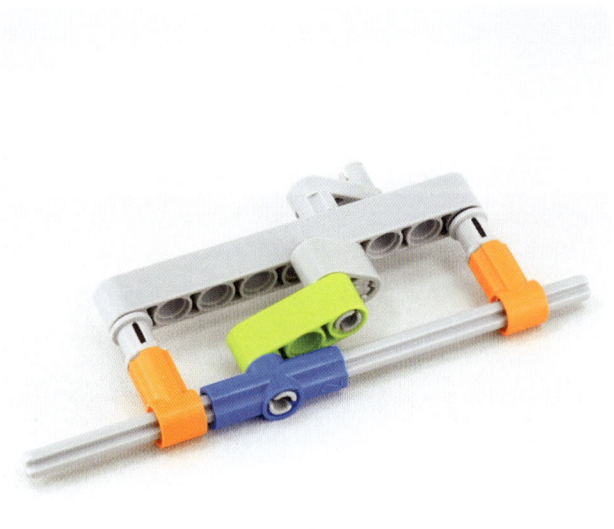

#83

×6 ×2

×2 ×2 ×3 ×2 ×2

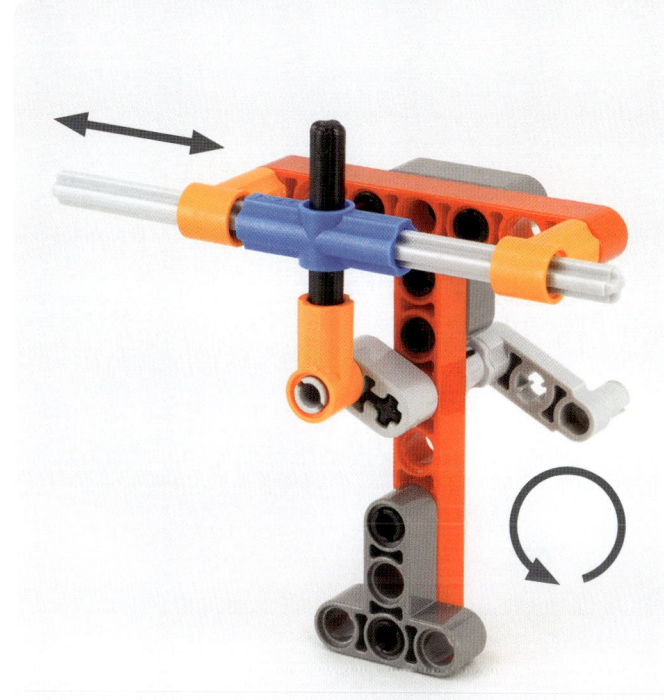

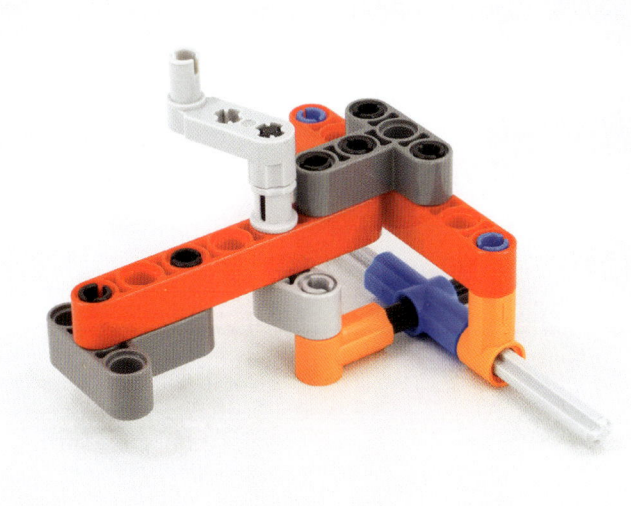

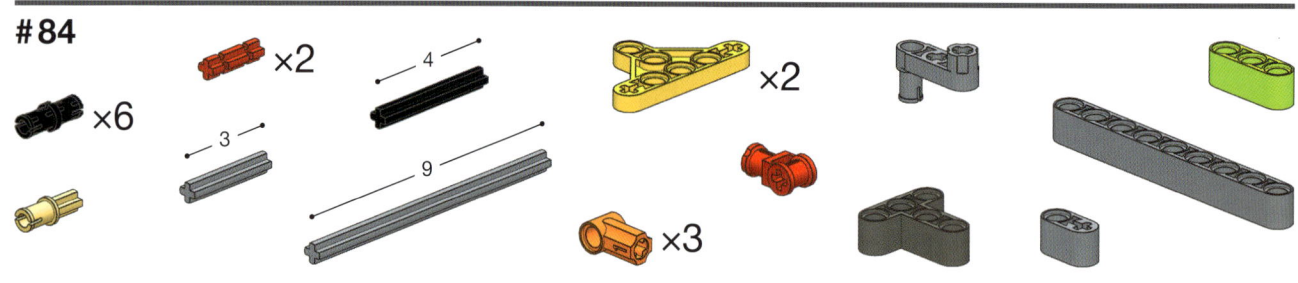

×6

×2

— 3 —

— 4 —

9

×2

×3

×2
×3
×2
4
×2
6 ×2
×4
×2

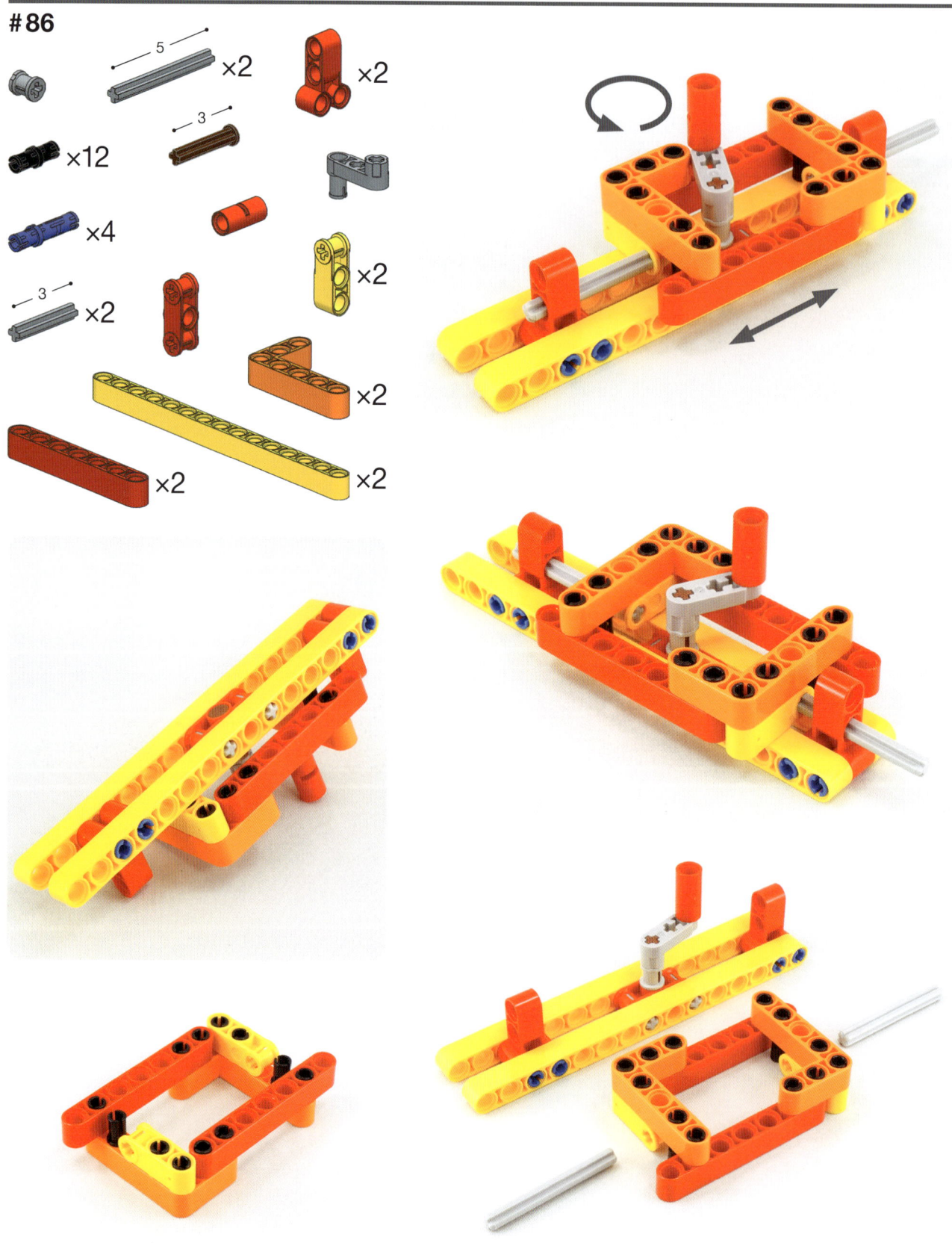

#87

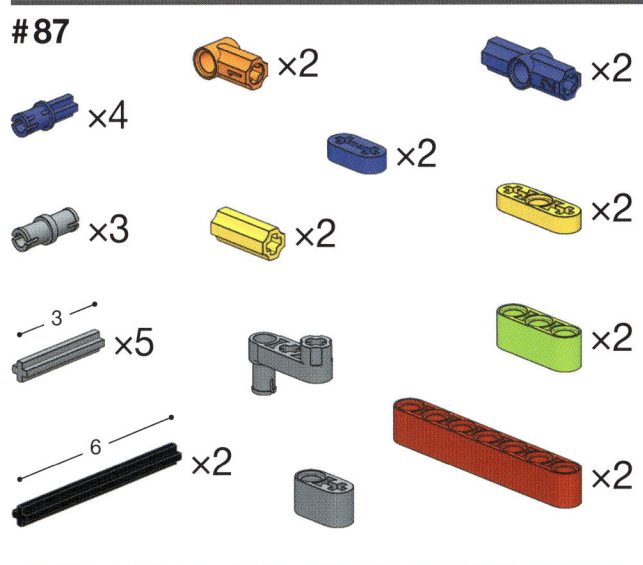

×4 ×2 ×2

×2

×3 ×2 ×2

3
×5 ×2

6
×2 ×2

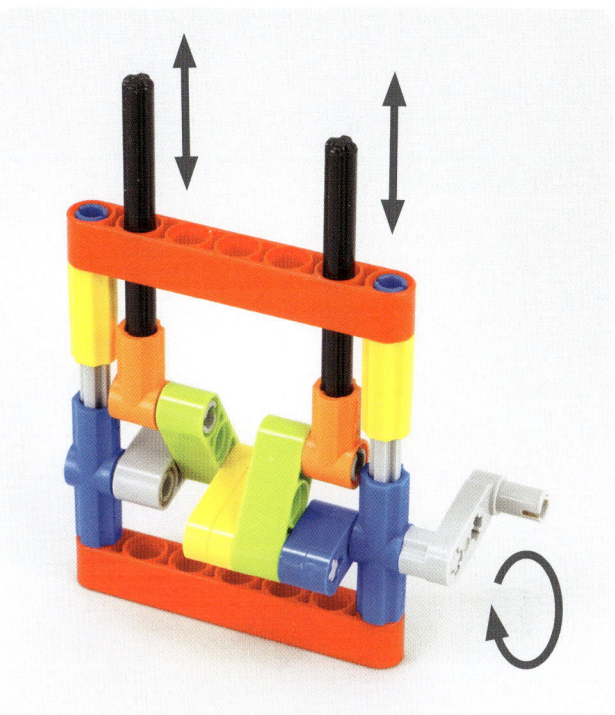

Rack and pinion gears

#88

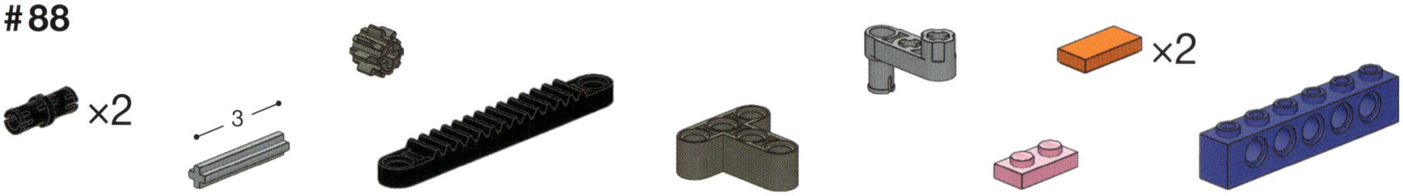

×2   3   ×2

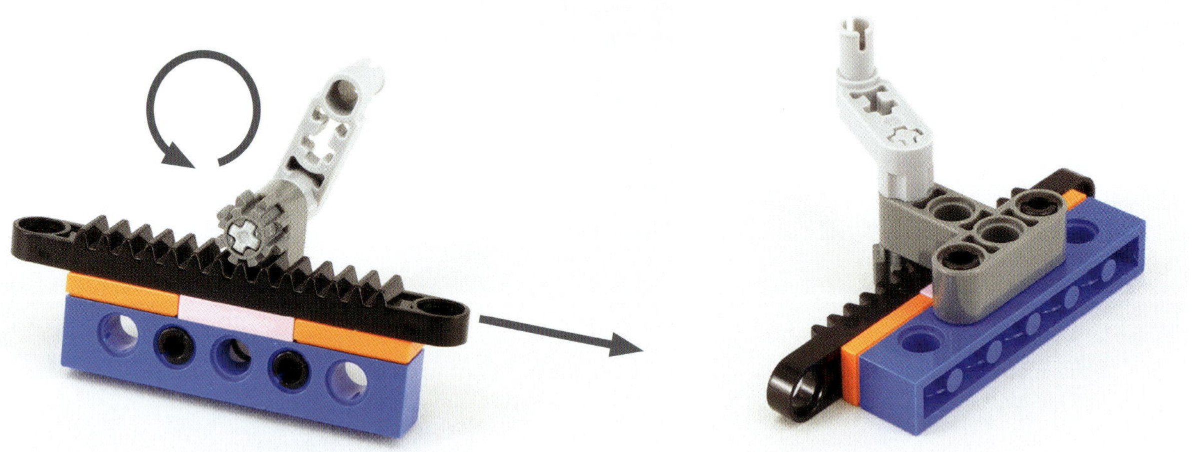

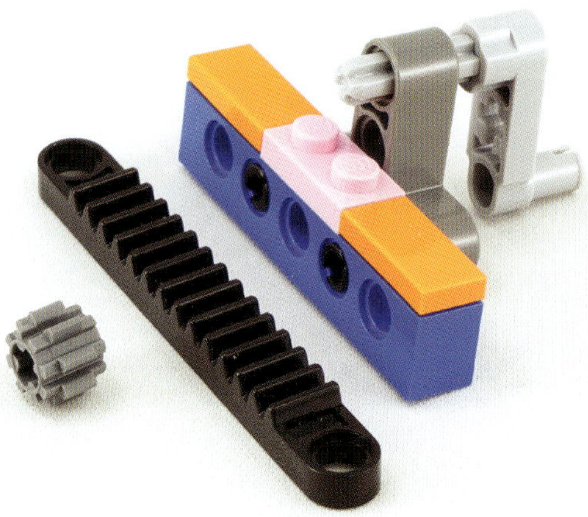

#89

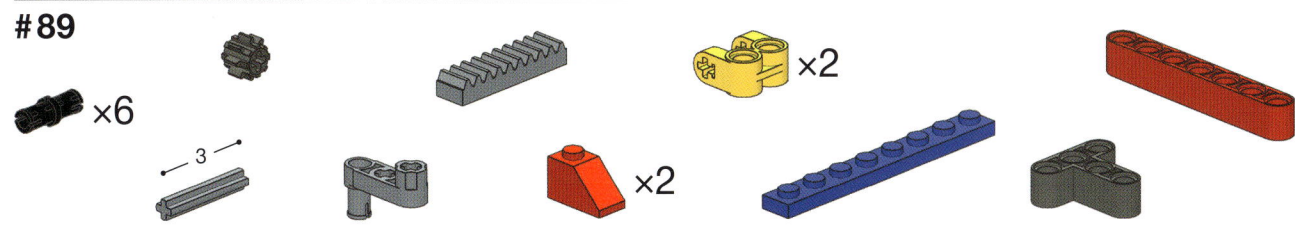

×6

3

×2

×2

#90

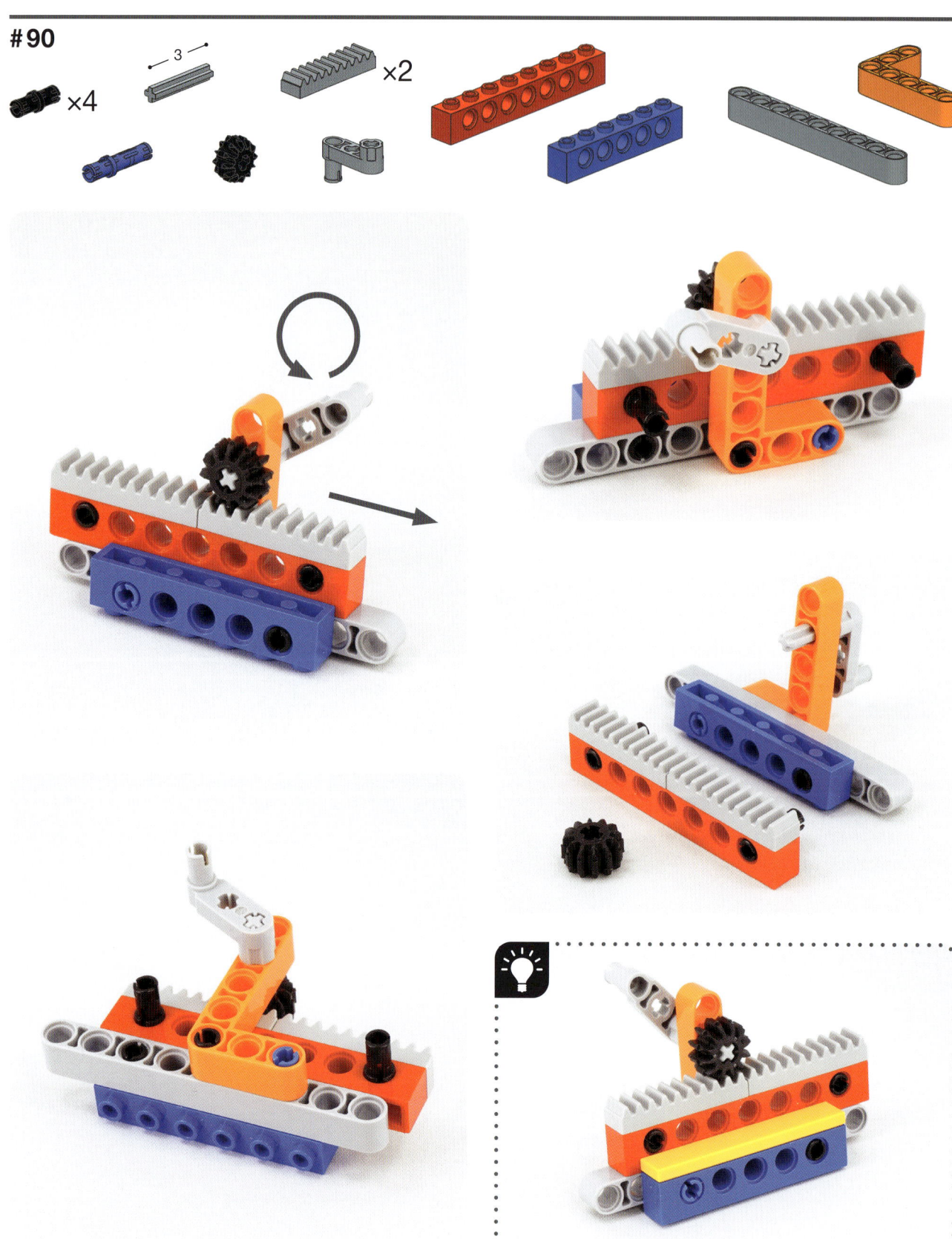

×4 — 3 — ×2

#91

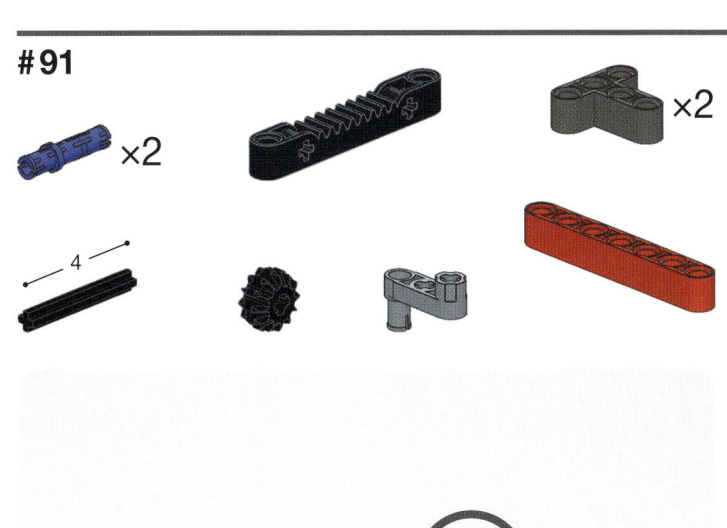

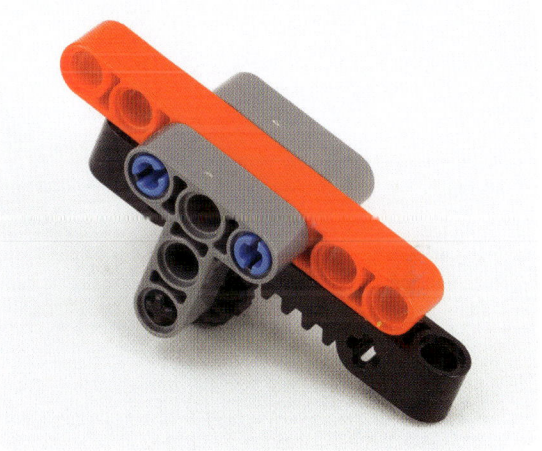

#92

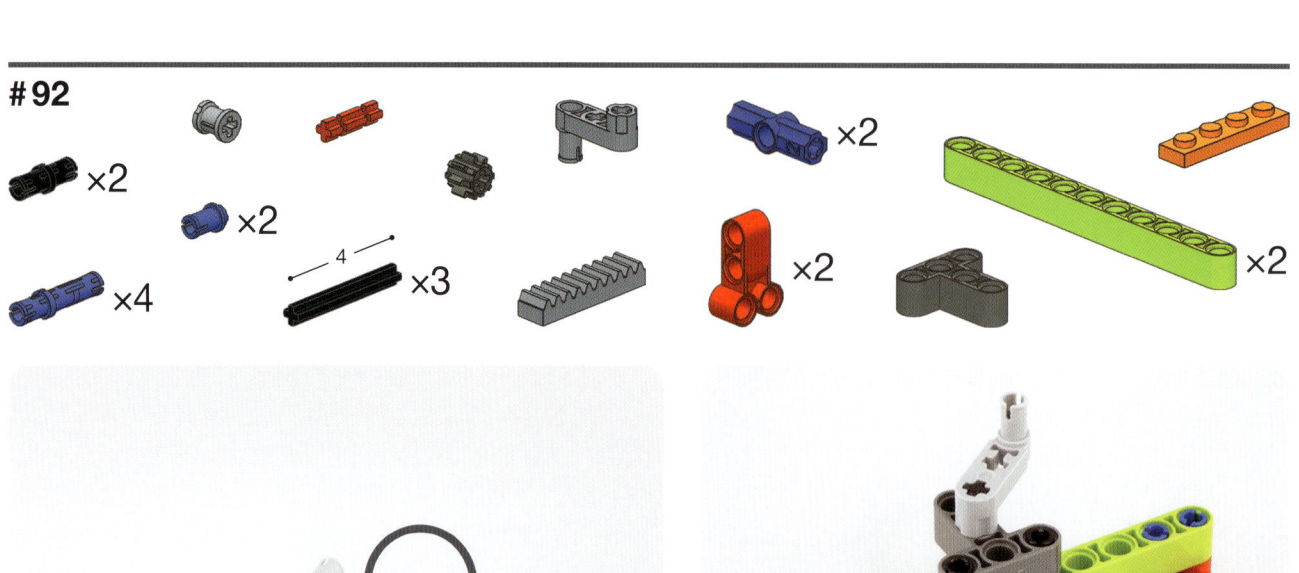

×2 ×2 ×2

×4 ×3 ×2 ×2

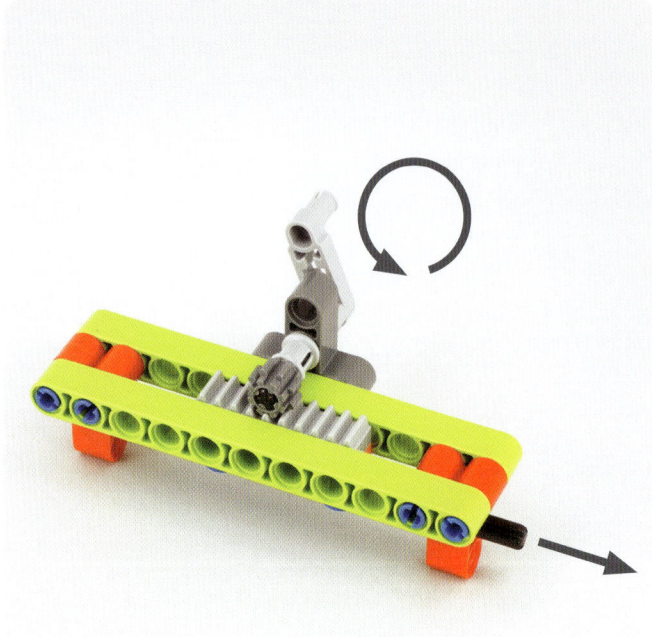

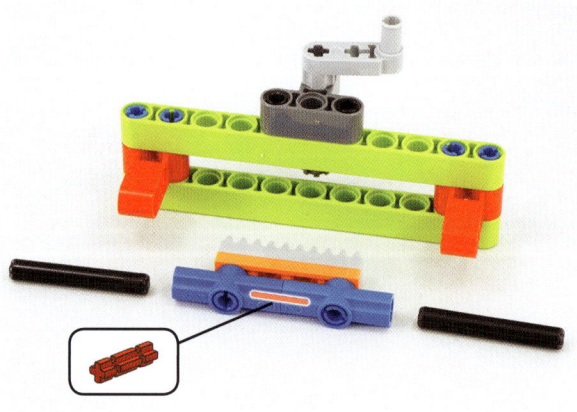

#93

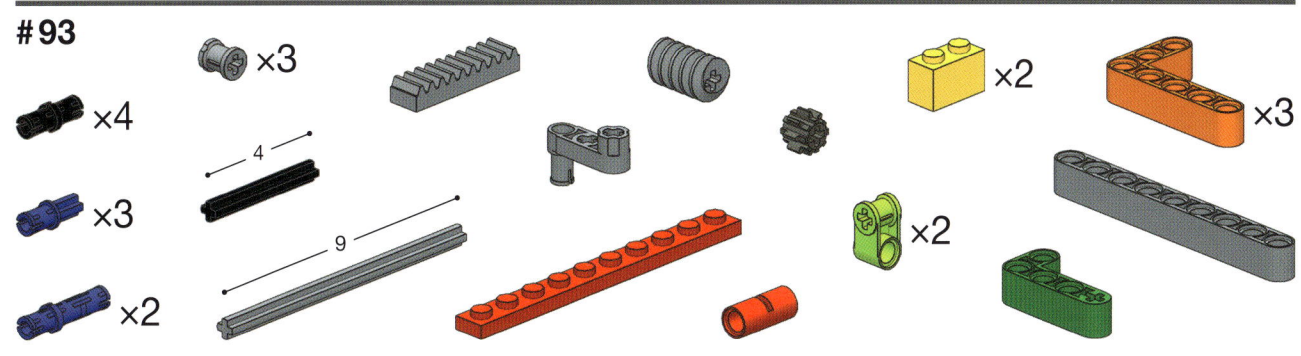

×4
×3
×2
×3
4
9
×2
×3

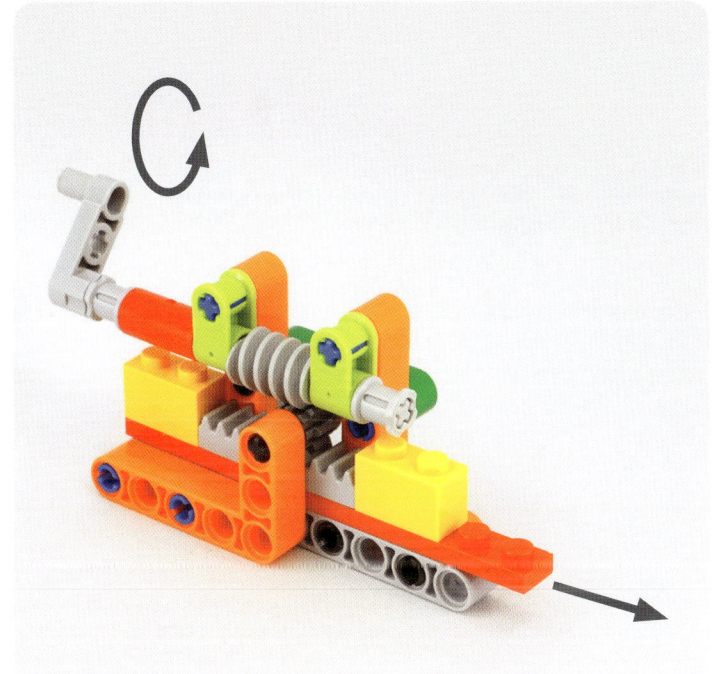

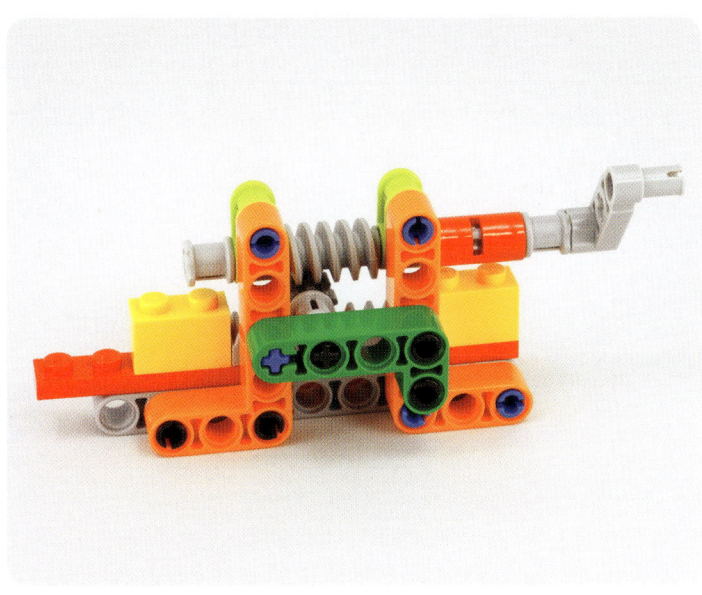

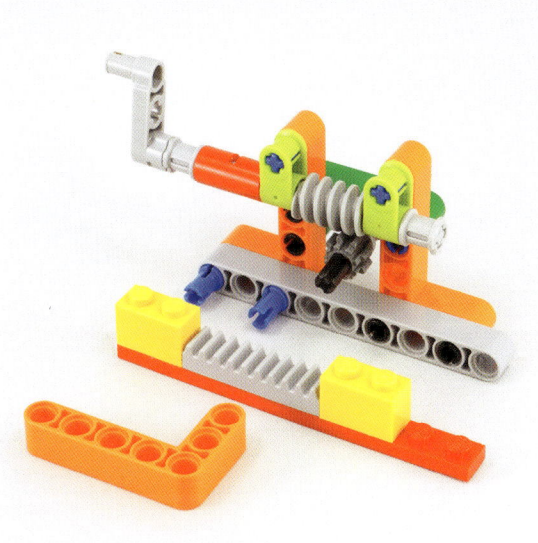

#94

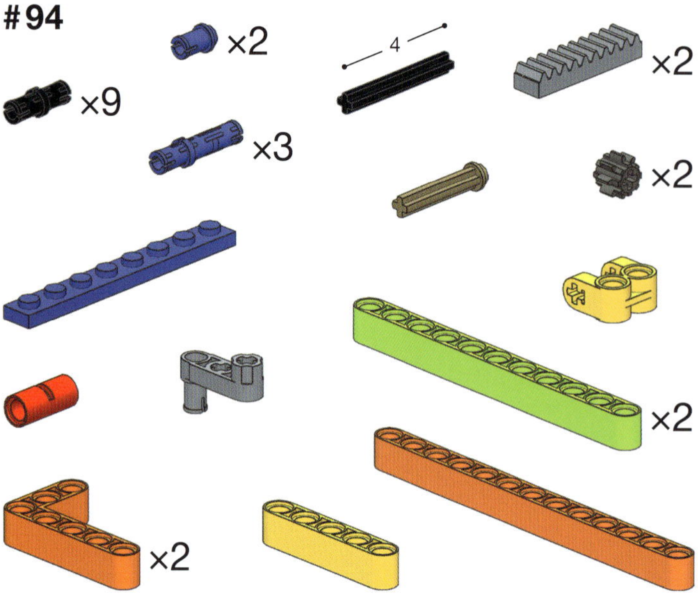

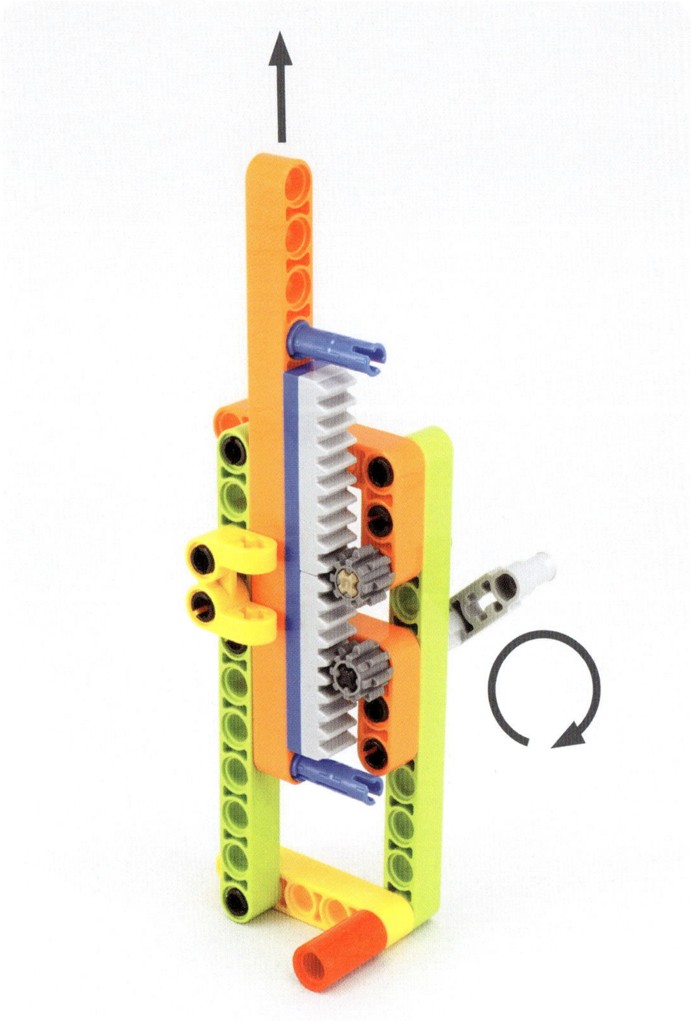

×9 ×2 ×3 4 ×2 ×2 ×2 ×2

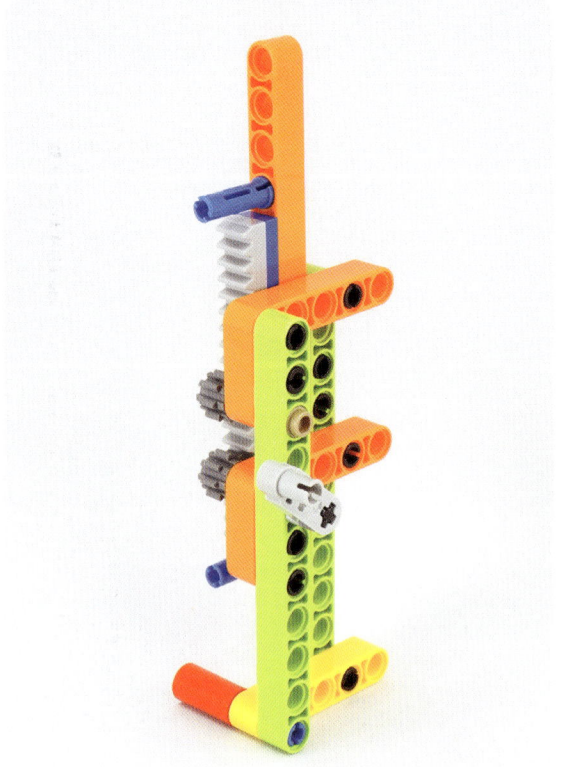

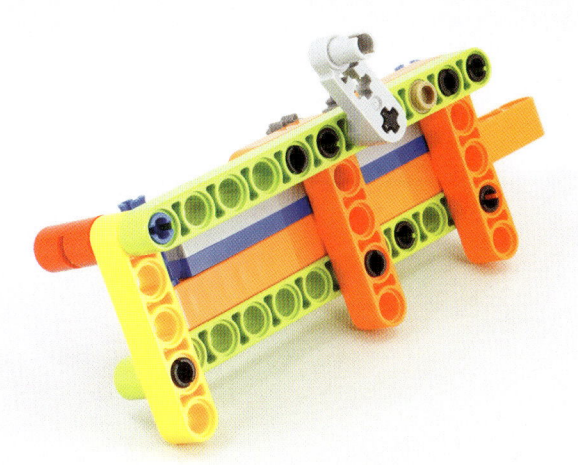

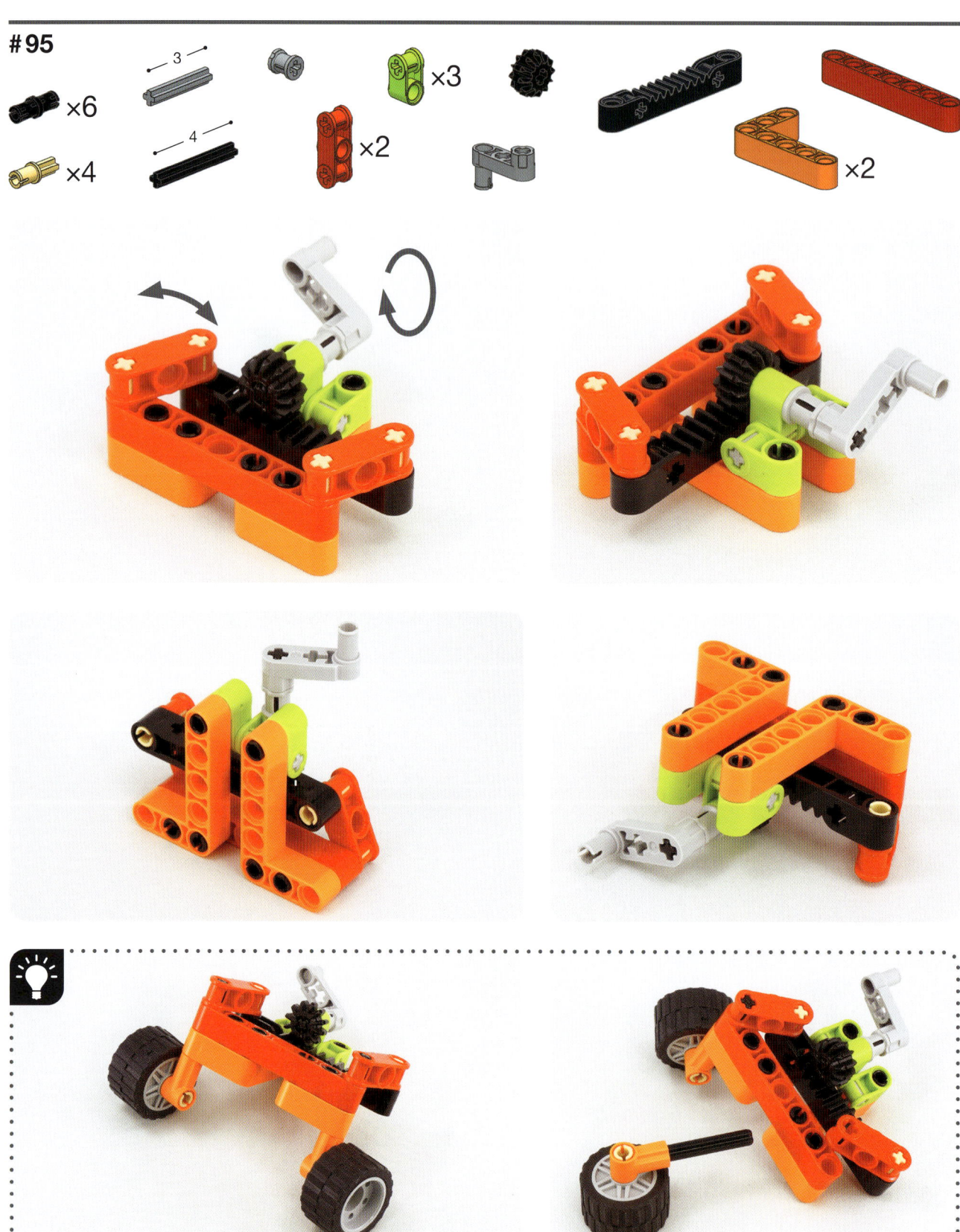

#95

×6
×4
×3
×2
×2

#96

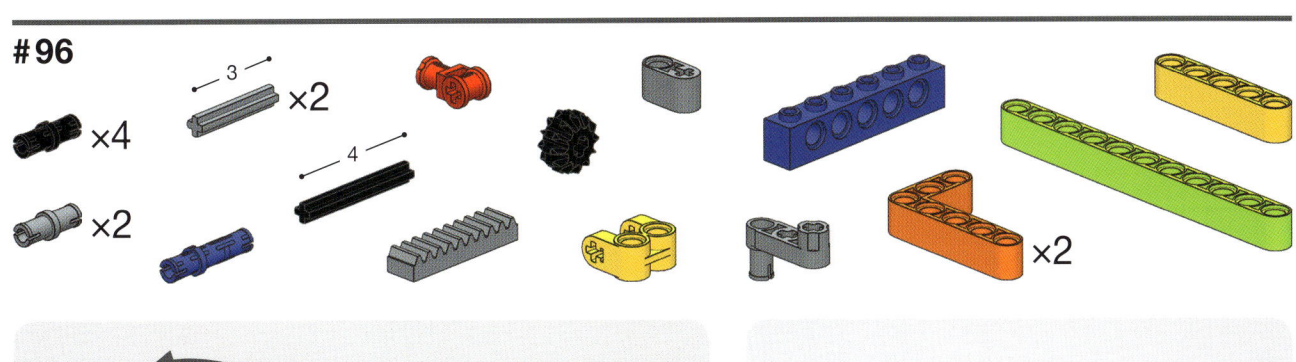

×4 ×2 ×2

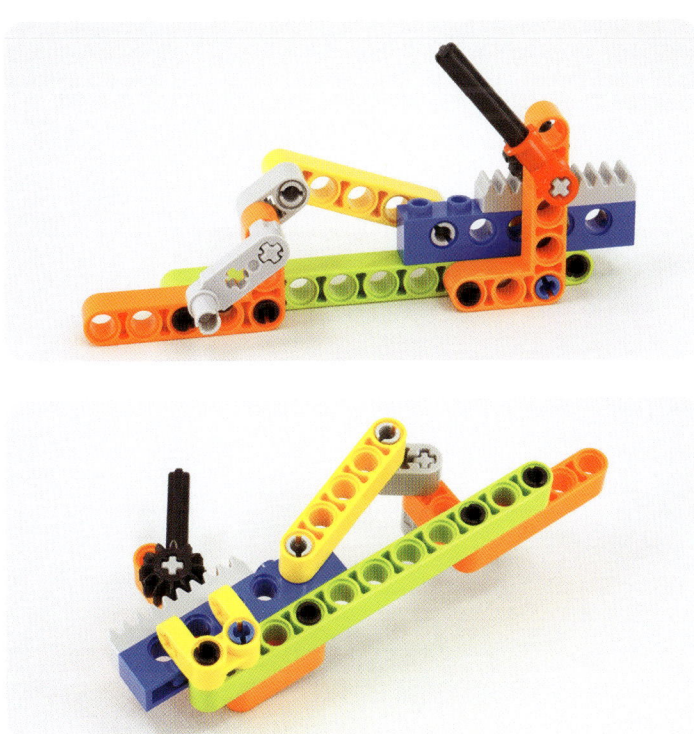

PART 2
Moving Vehicles

Simple cars

#97

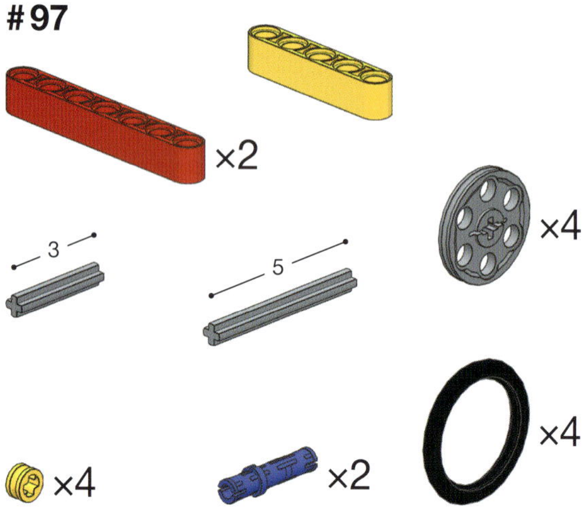

×2

3

5

×4

×4

×4

×2

#98

#99

×4
×4
×2 ×4
×3 ×2

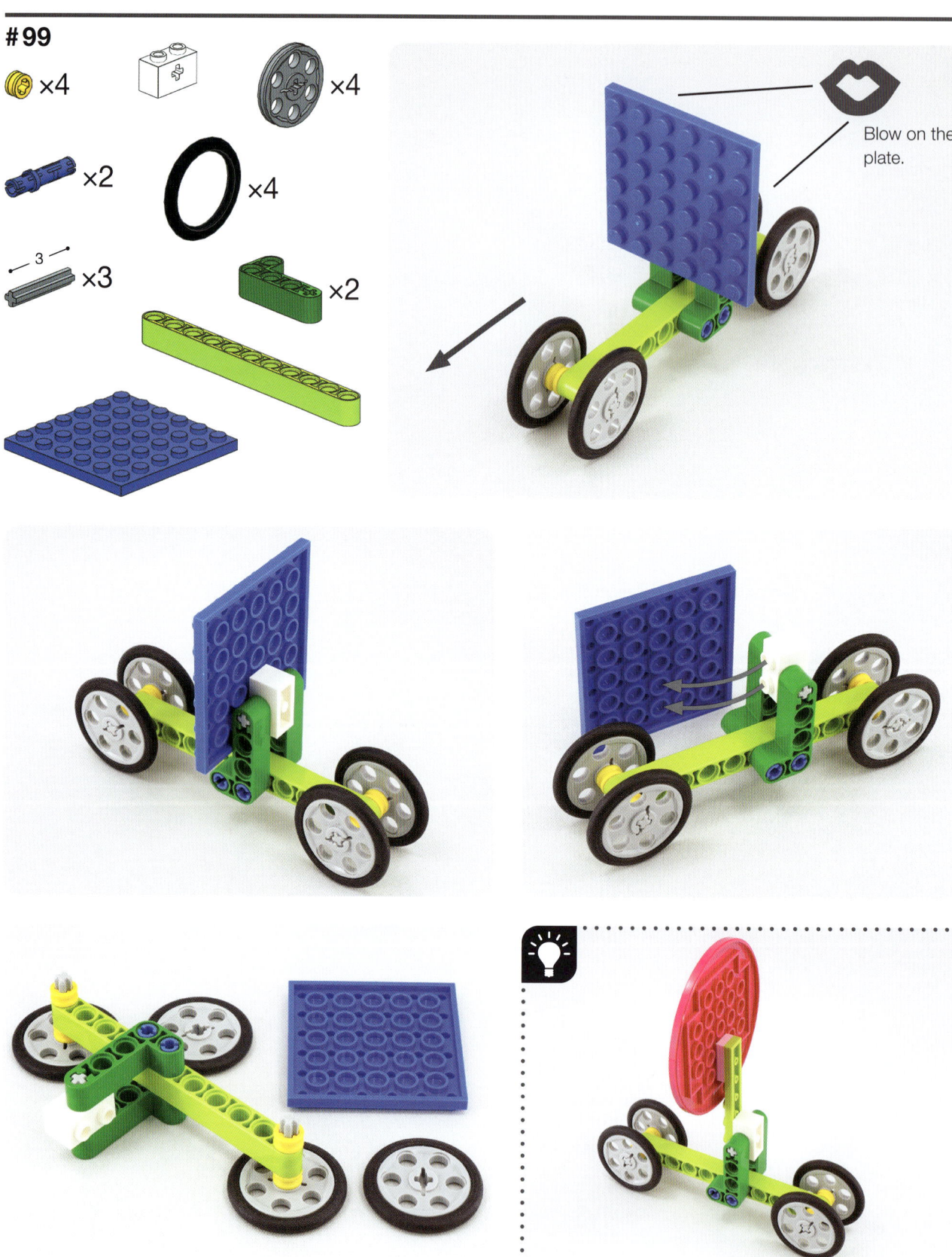

Blow on the plate.

#100

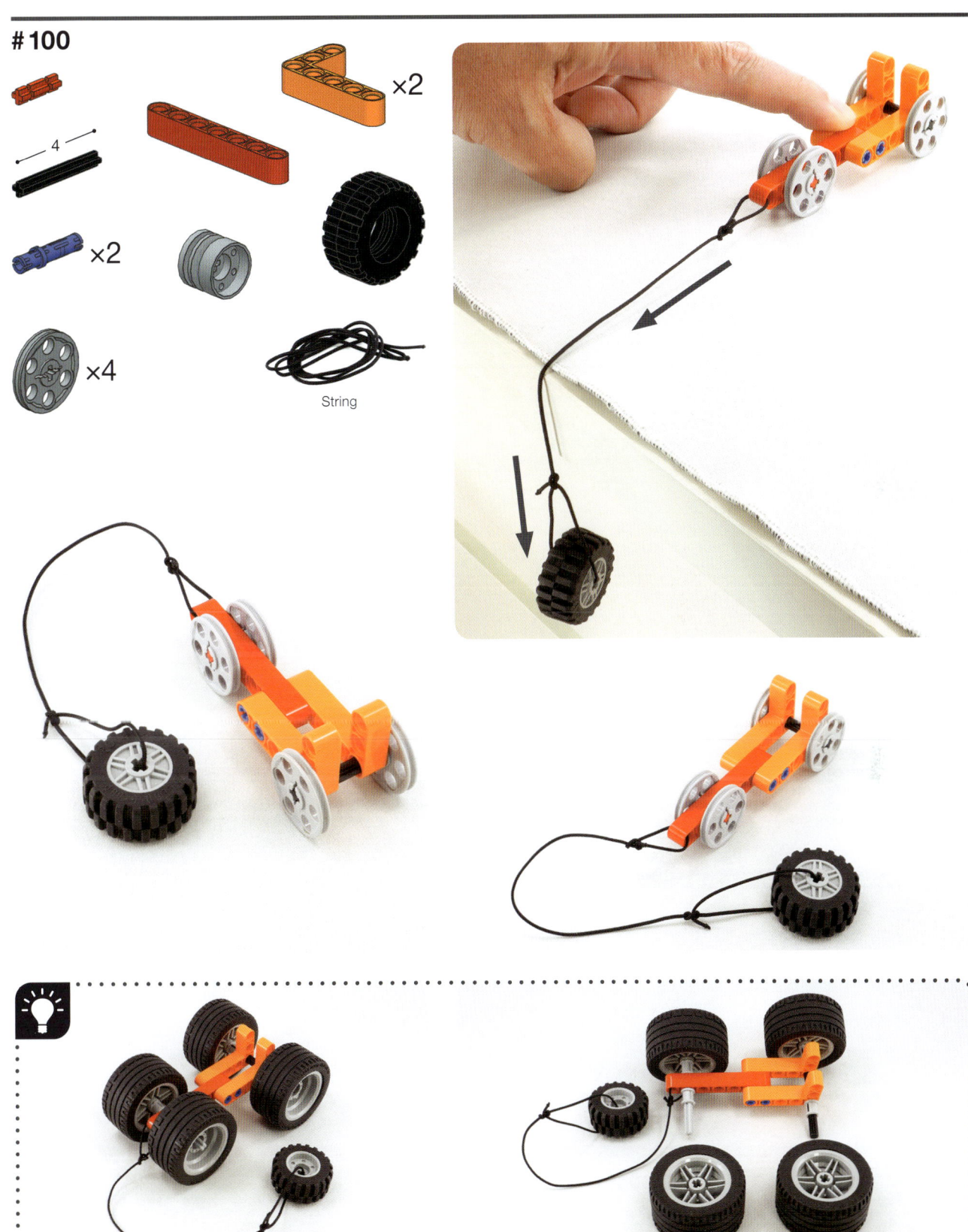

String

Cars with rotary motion

 # 101

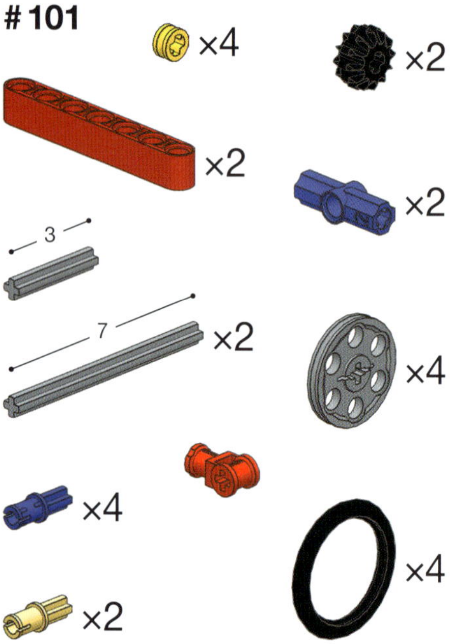

×4   ×2

×2

3

7   ×2

×4

×4

×4

×2

#102

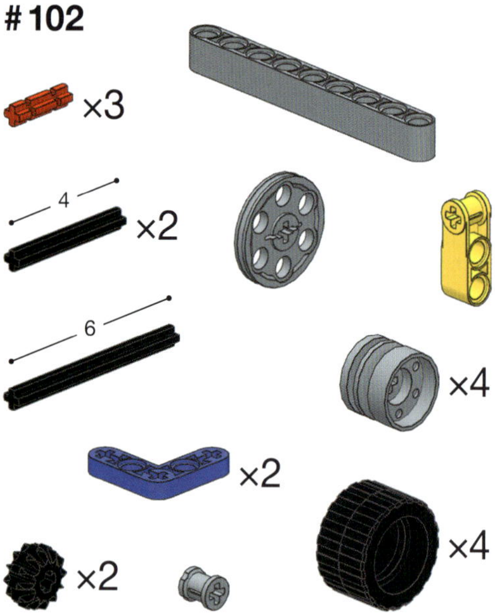

×3

4 ×2

6

×4

×2

×4

×2

#104

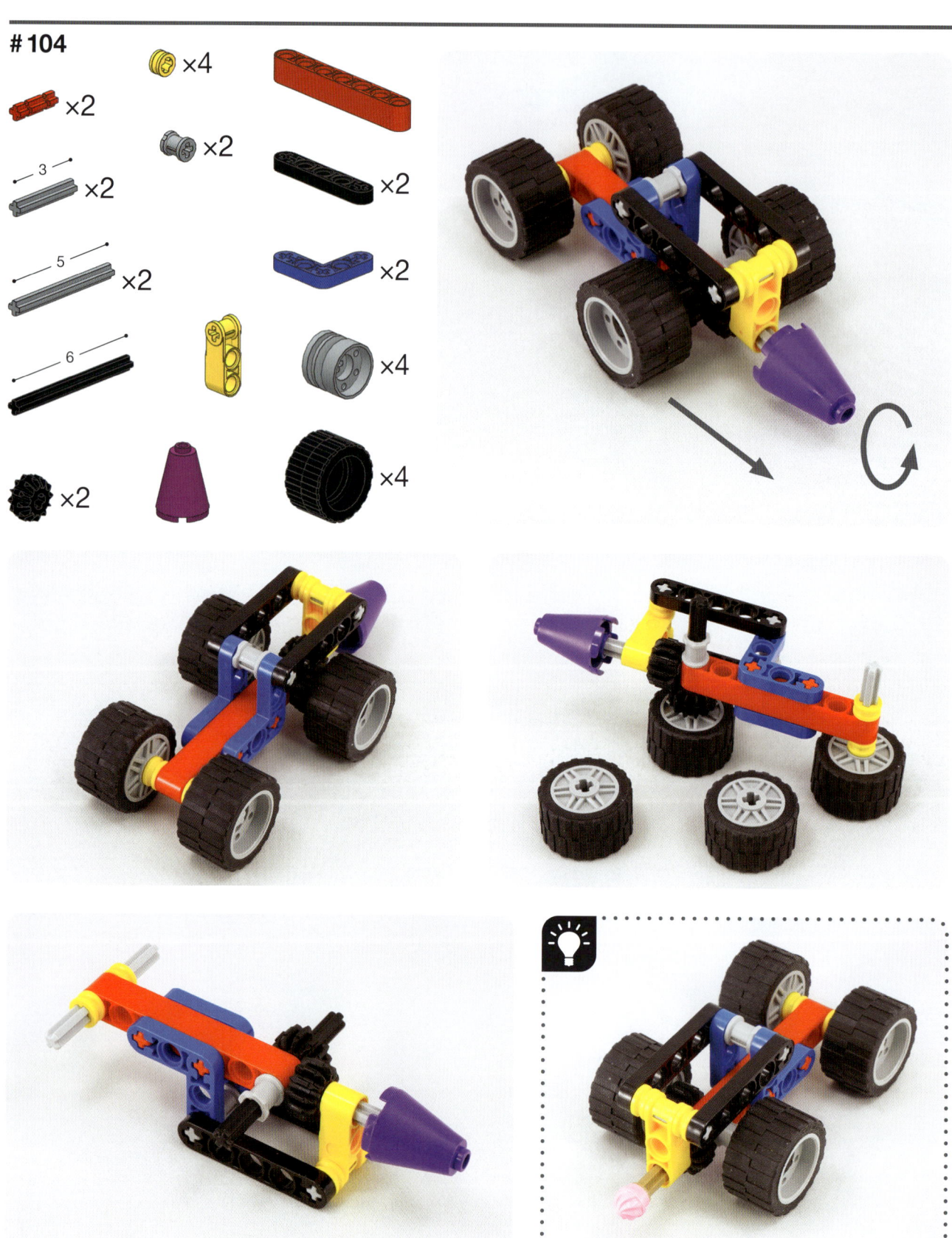

×4

×2

3 ×2

5 ×2

6

×2

×2

×2

×2

×4

×4

#105

×6

×2

×4

×2

×2

×2

×2

3

6

8

×2

×2

×4

×4

#107

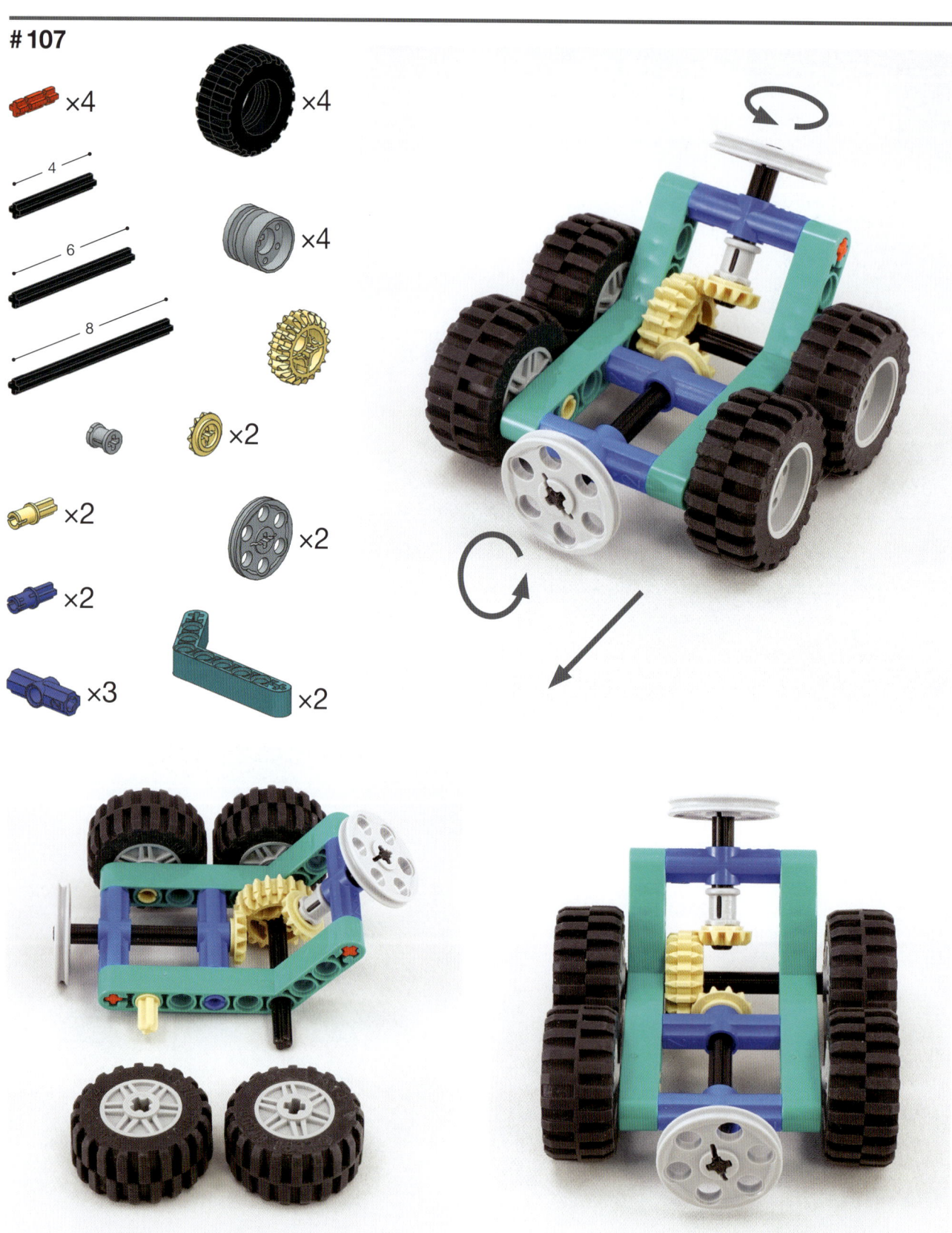

×4

×4

4

6

×4

8

×2

×2

×2

×2

×3

×2

104

#108

Blow on
the plate.

#110

×8 ×2 ×2 ×2 ×2 ×4

3 ×3 ×2 ×2 ×4

5 10 ×2 ×2

Cars with oscillating motion

 # 111

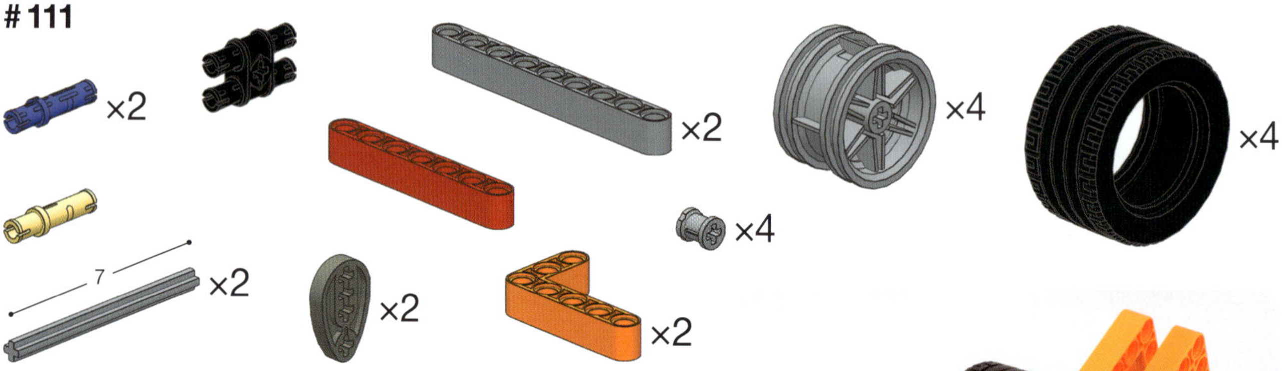

×2 ×2 ×4 ×4

×4

×2 ×2 ×4

×4　×2　×4　×2　×2　×4　×4

113

×3

×2

×2

×2

×2

×2

×2

×2

×4

×4

×4
6
×2
×2
×3

×6
7
×2
×2
×3

×2

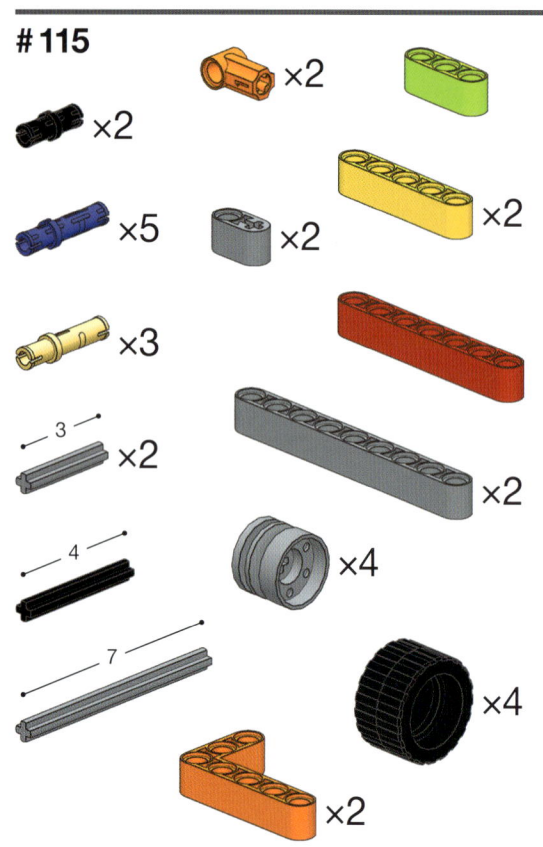

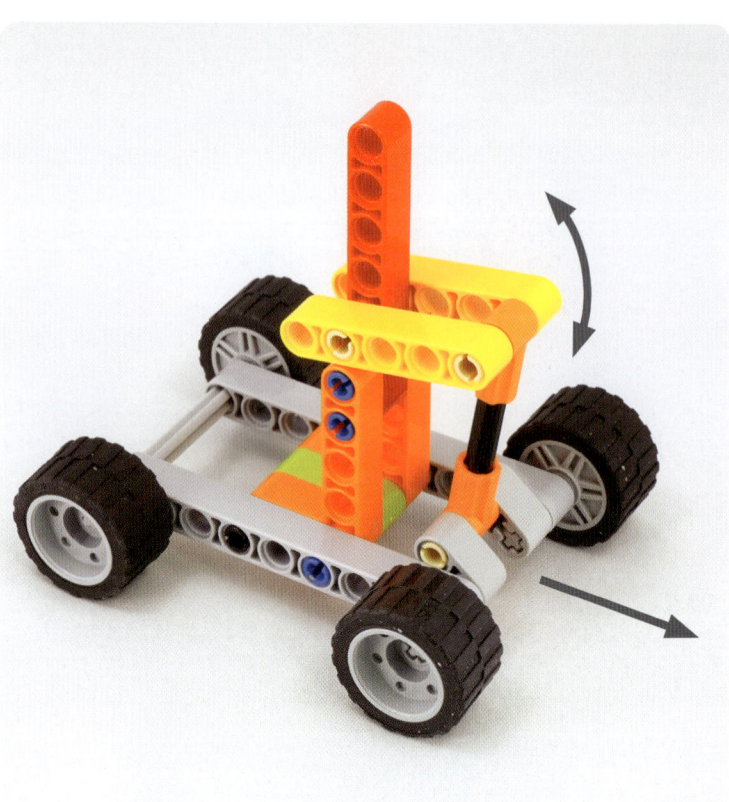

×4

×2

×6

3 ×4

5 ×2

4

8

×4

×2

×4

×4

×4

×2

×4

×4

×4

#118

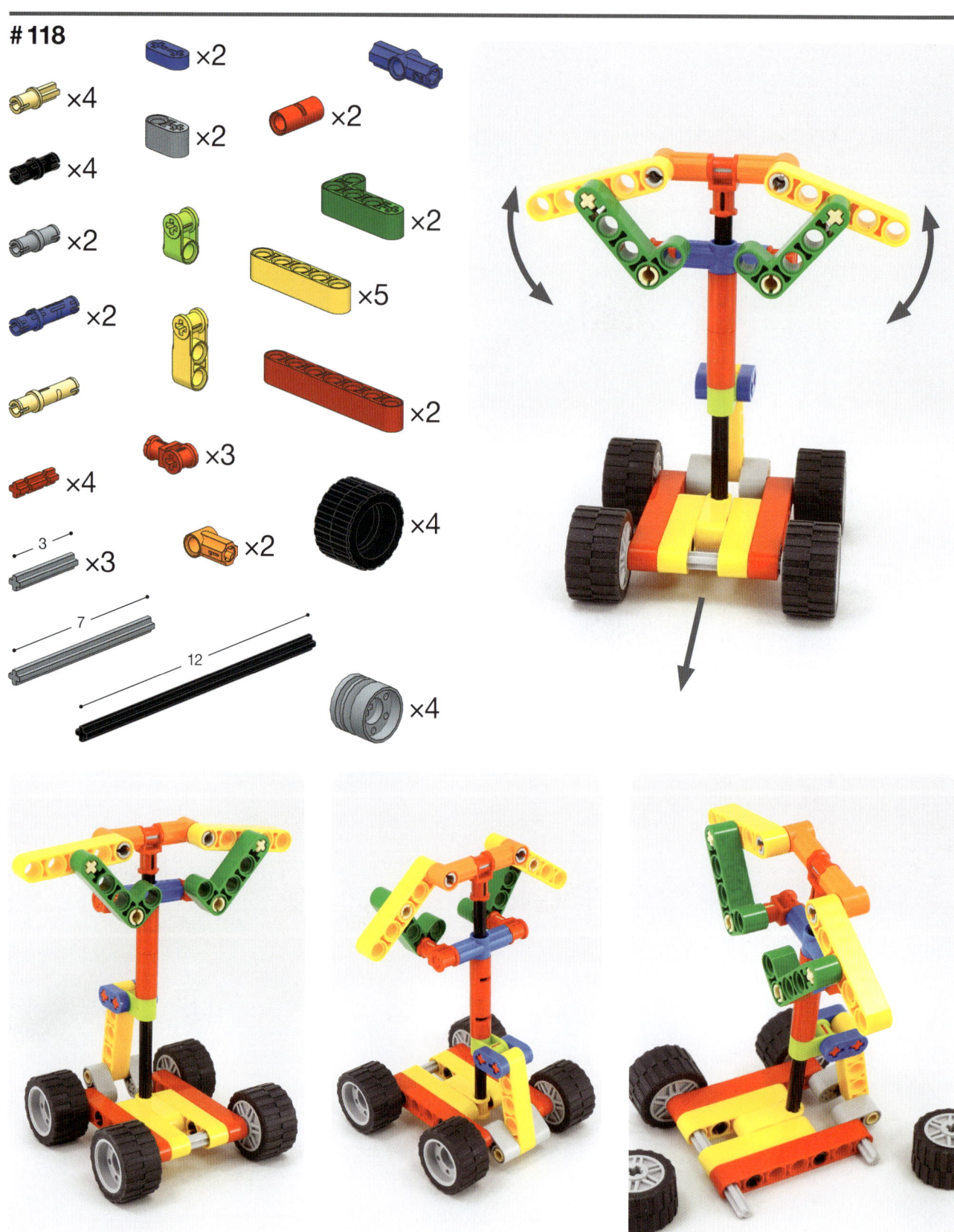

Cars with reciprocating motion

 # 119

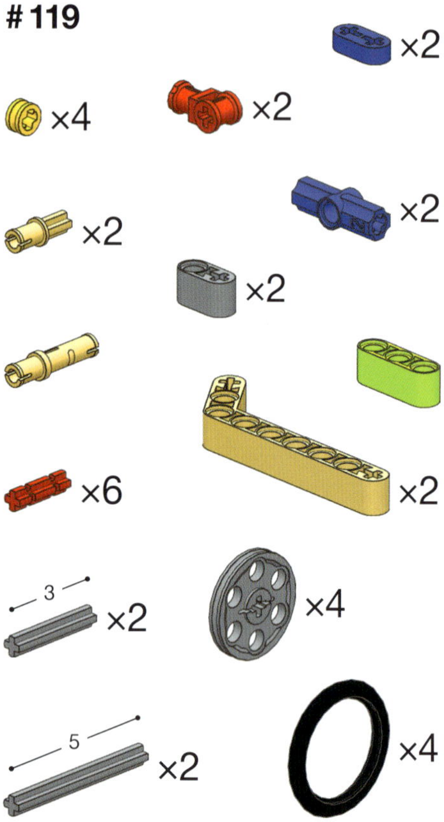

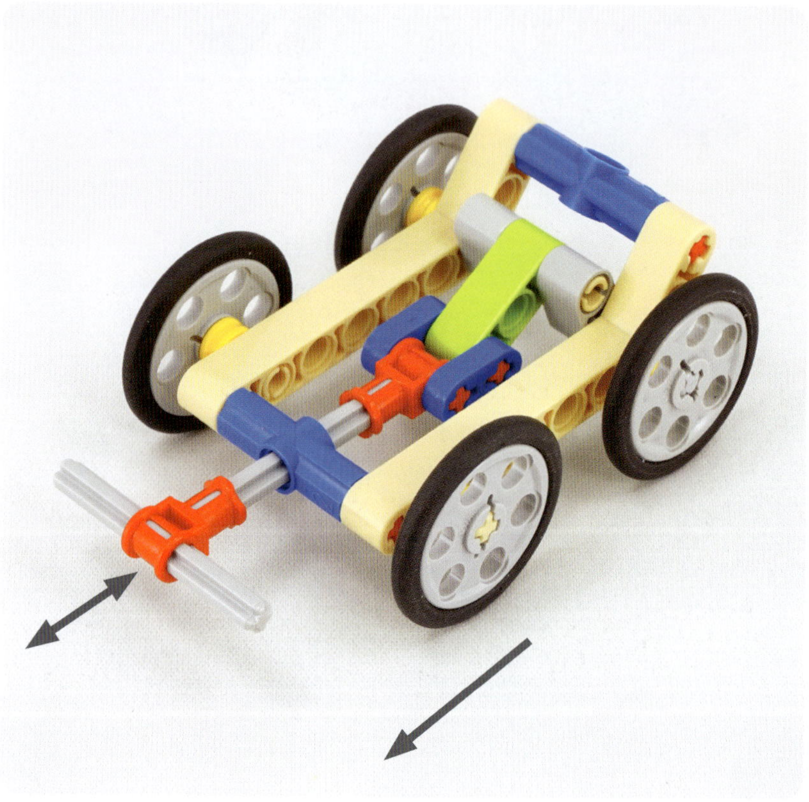

×2

×4   ×2

×2   ×2

×2

×2

×6   ×2

×3 ×2   ×4

×5 ×2   ×4

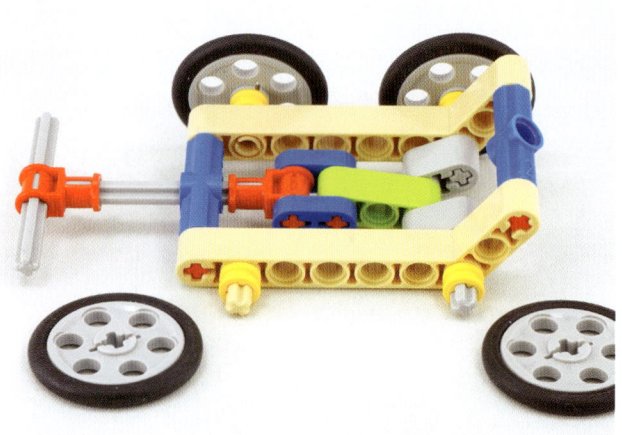

×4 ×2 ×2

×2 ×4 5

×2 ×4 7 ×2

×2 ×2 ×2 ×4

4 ×2 ×2 ×4

×2

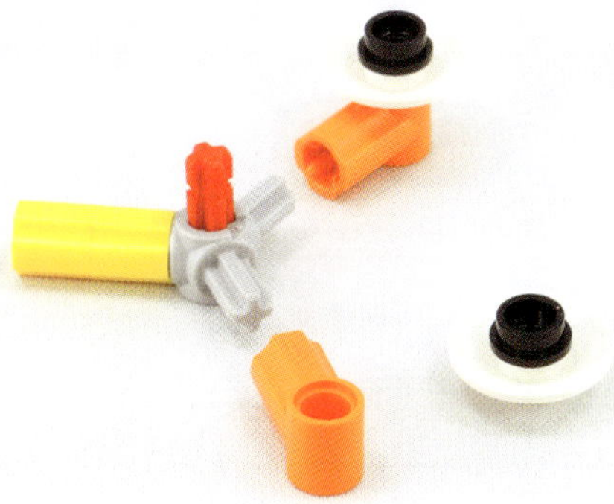

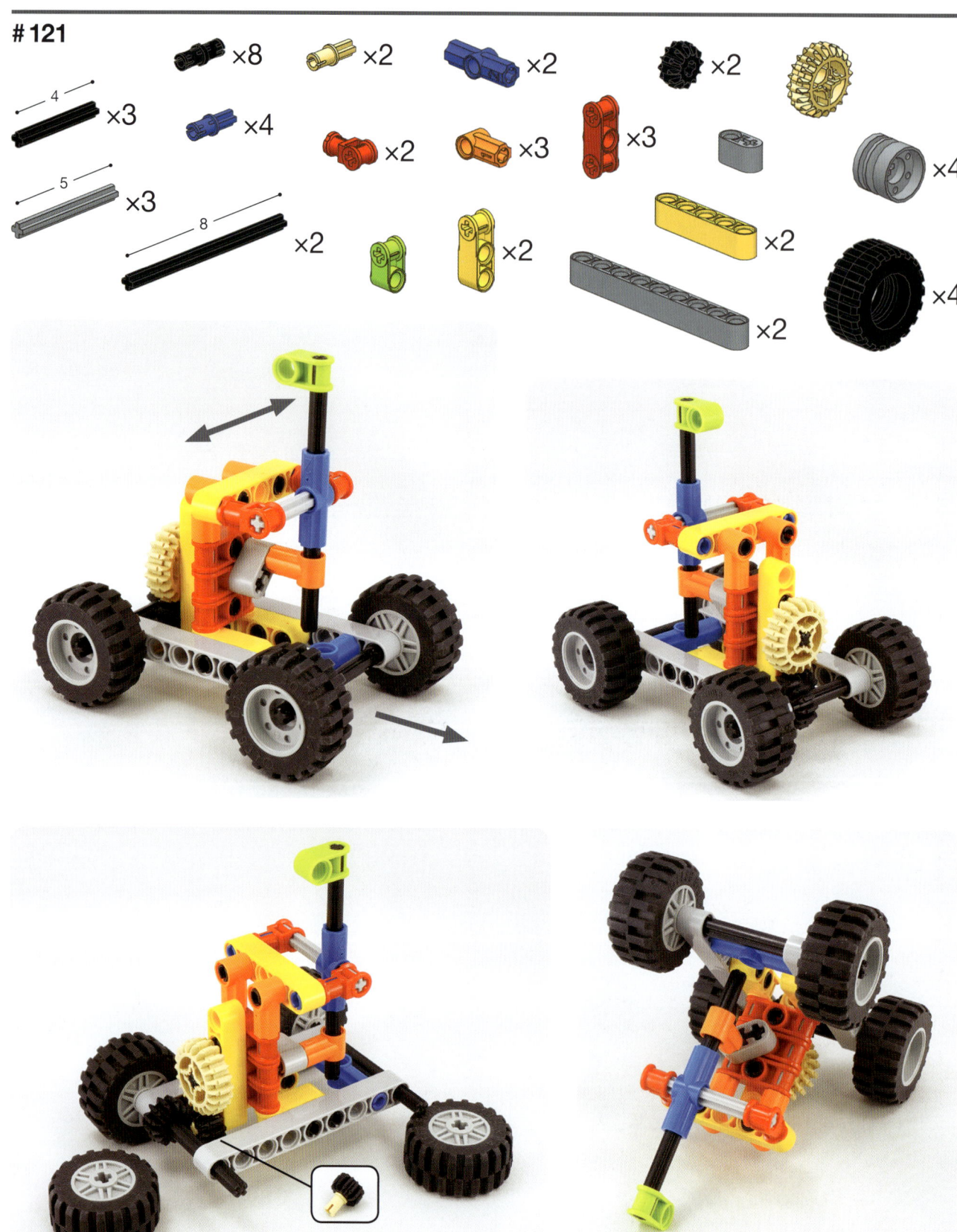

#121

Cars that run on energy stored in the flywheel

 # 122

×2

6

×2

7

×2

×2

×2

3 ×3

×2

×2

×2

4

×4

×4

×2

When you push the car with your hand, energy is stored in the flywheel, and when you let go, the car runs on that energy.

Flywheel

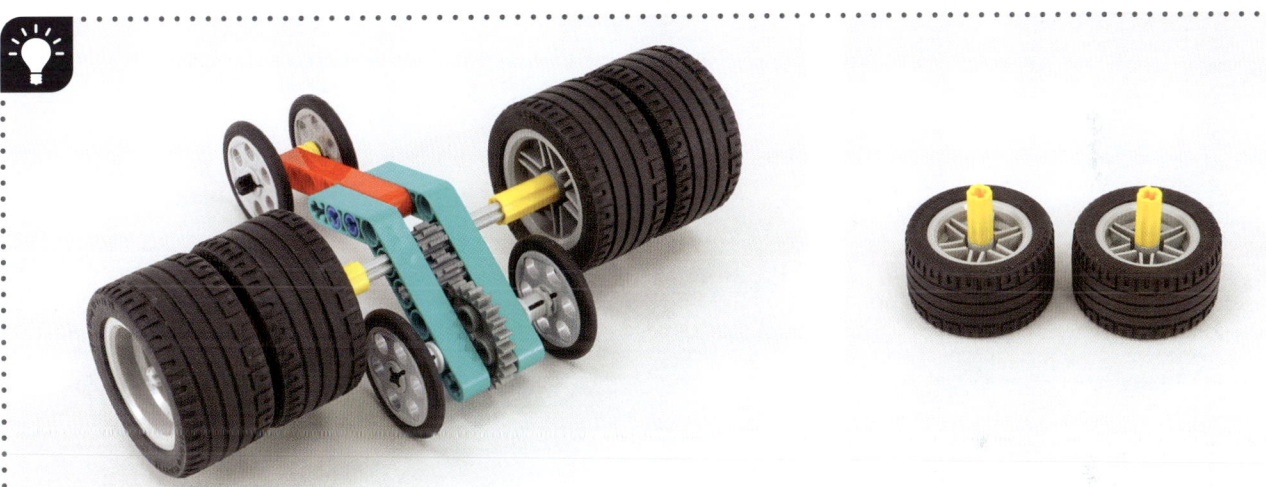

#123

#124

125

Cars that run using weights

×9 ×2 ×2 ×5 ×2 ×4 ×2 ×2

5
6
7
8
10

×2 ×2 ×4 ×4 ×2

When you lift the weight and then let it go, the car starts running.

The rear wheel leaves the ground when the weight lands, and the car keeps running.

127

×2
7 ×2
×2

8 ×2
×4

×3
10 ×4

×13
×4

5 ×4
×2

×4
×4

×2
×2

×4

×4
×4

×2
×2 ×3

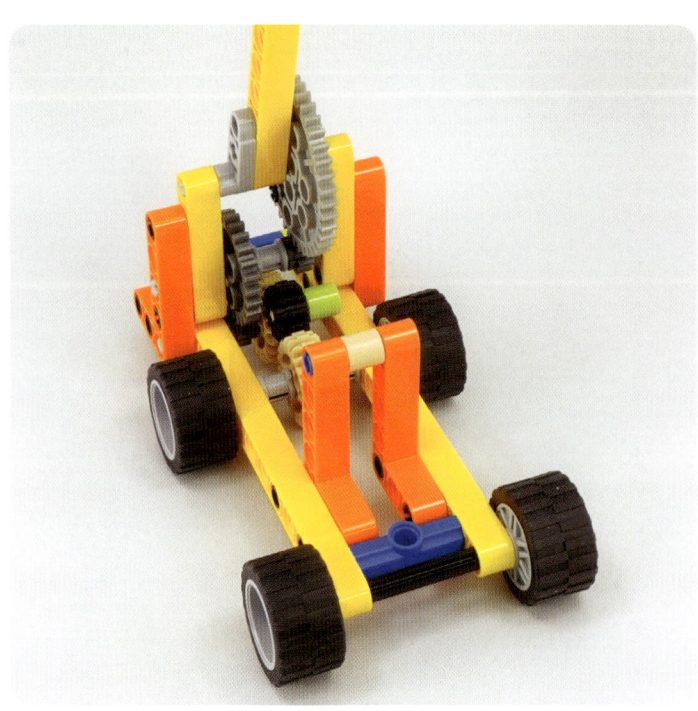

Cars that run using rubber bands

#128

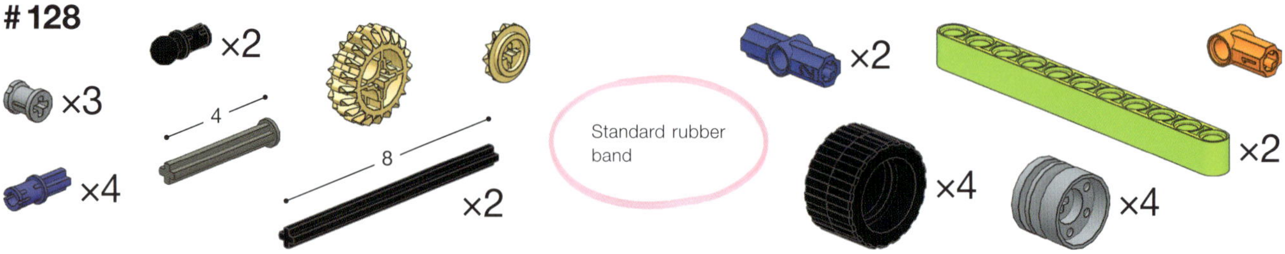

×2 ×3 ×4 4 8 ×2 Standard rubber band ×2 ×2 ×4 ×4

#129

Standard rubber band

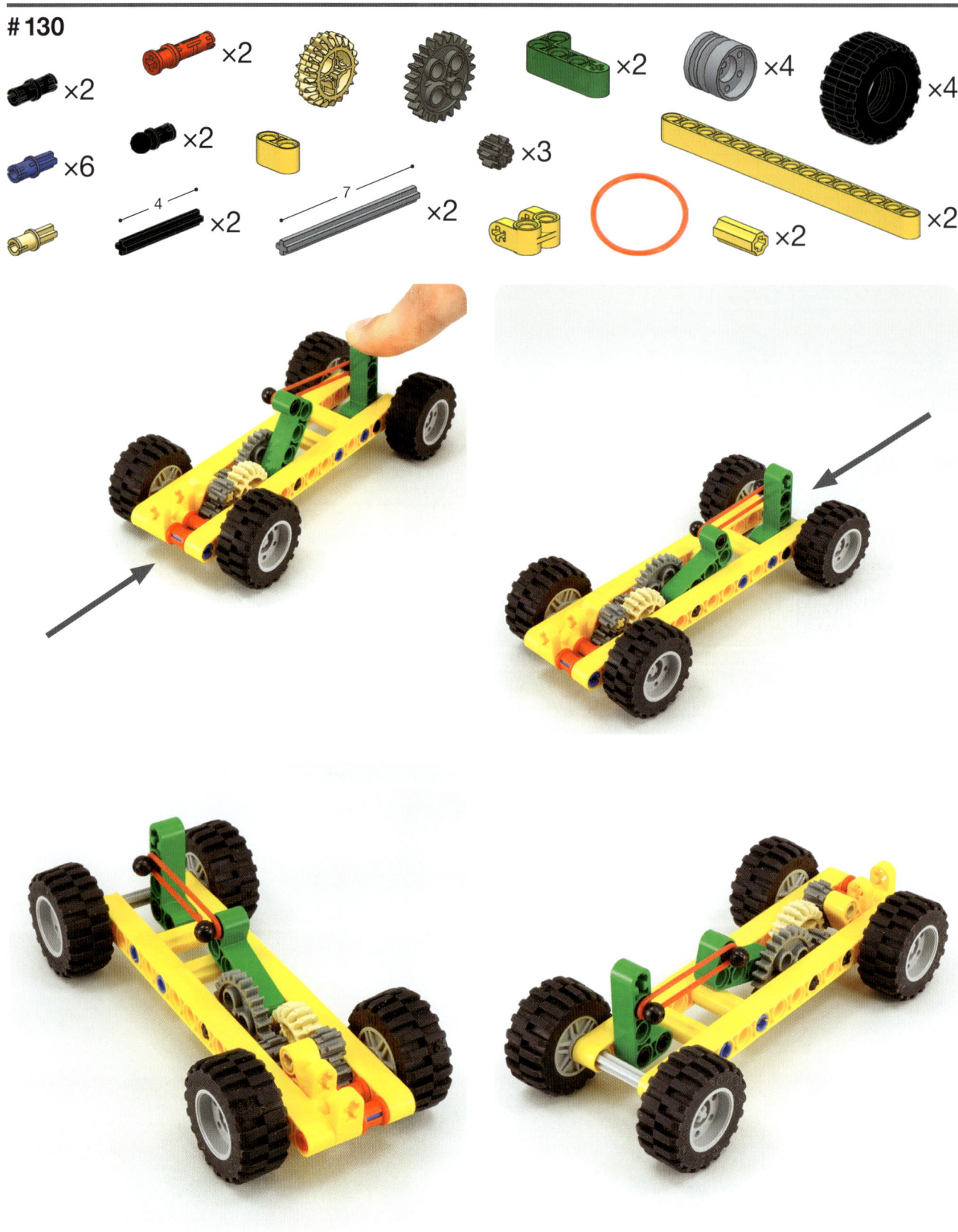

×2 ×2 ×2 ×4 ×4

×2 ×6 ×3

4 7 ×2 ×2 ×2

131

Standard rubber band

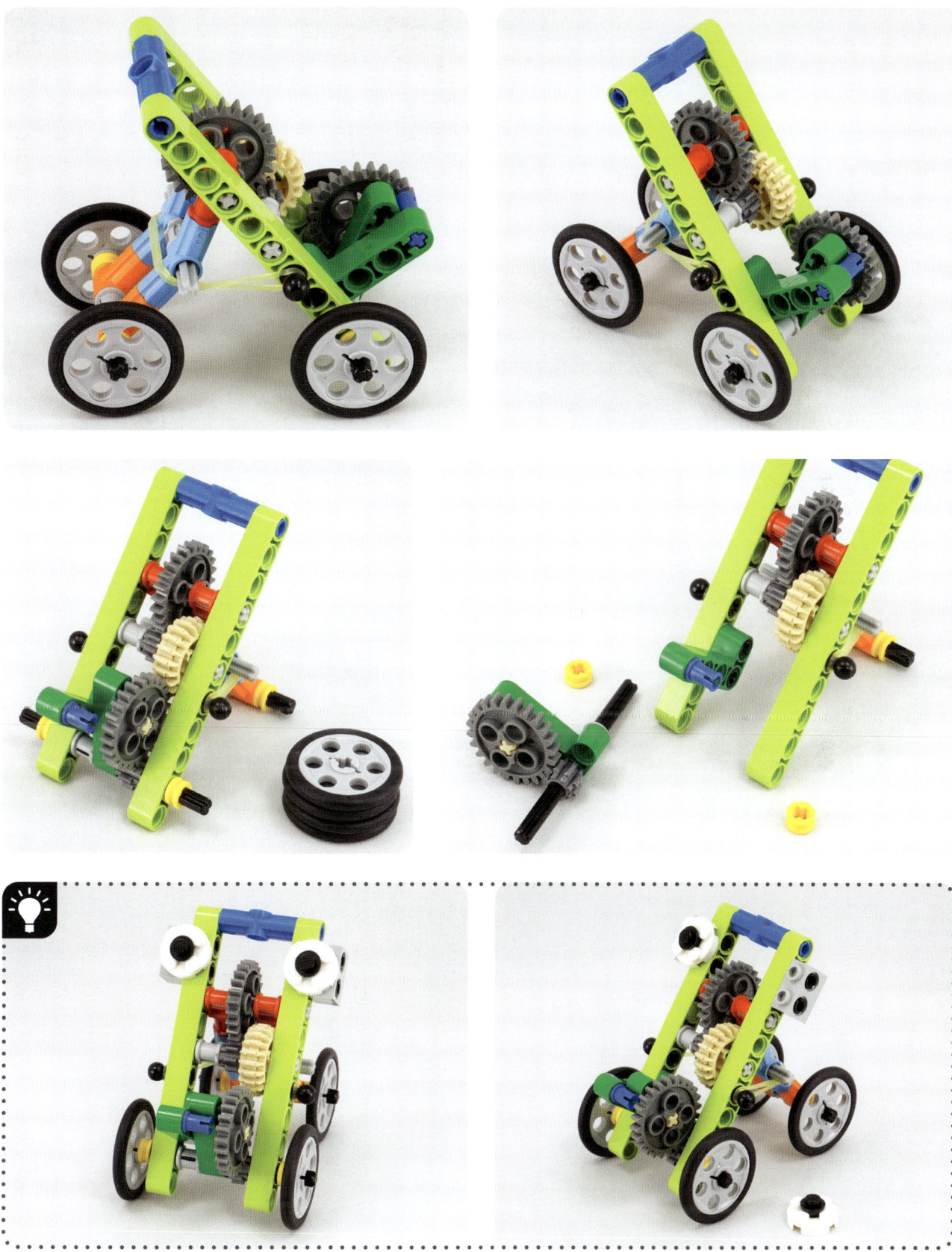

132

Standard rubber band

#133

#134

×9

×16

×2

×2

×2

×3

3 ×2

5

6

7 ×5

8

×4

×2

×6

×2

×2

×2

×4

Standard rubber band

×4

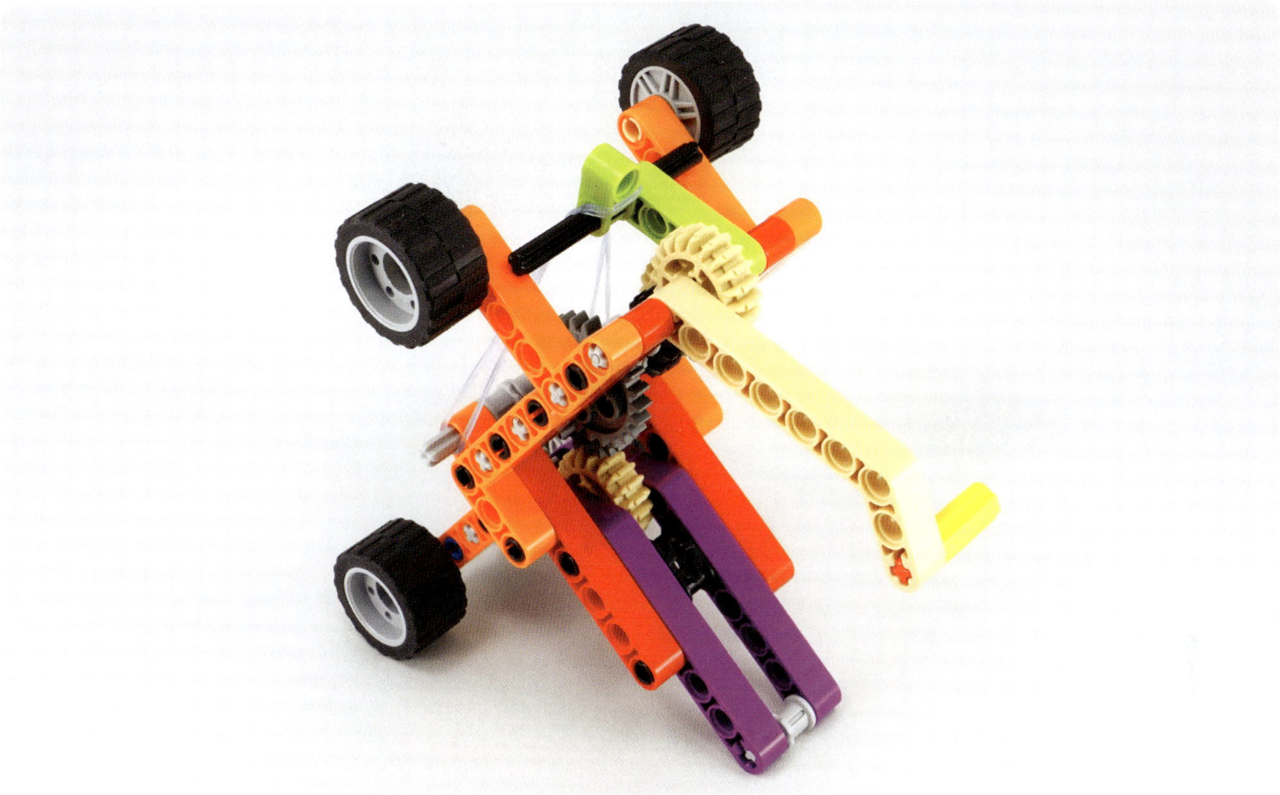

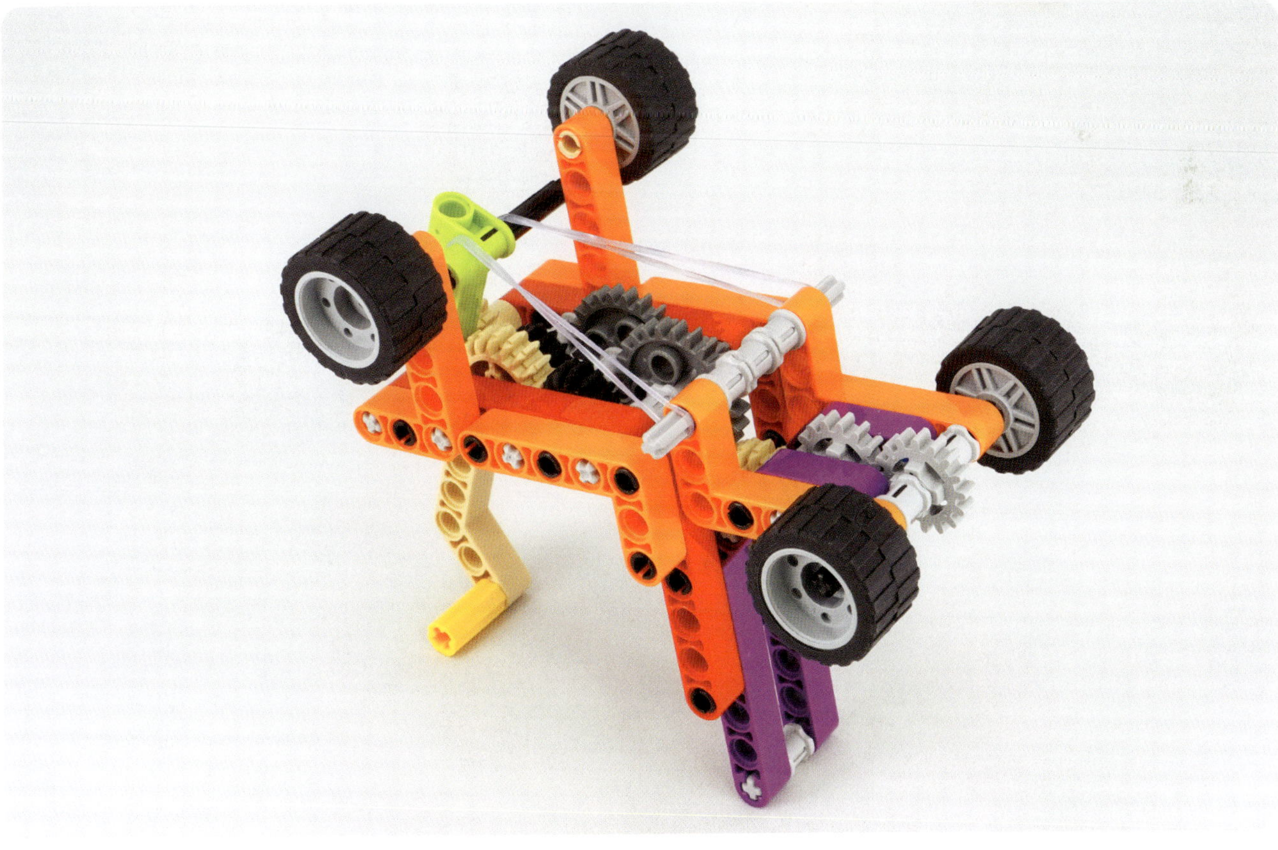

Cars that move forward and backward

#135

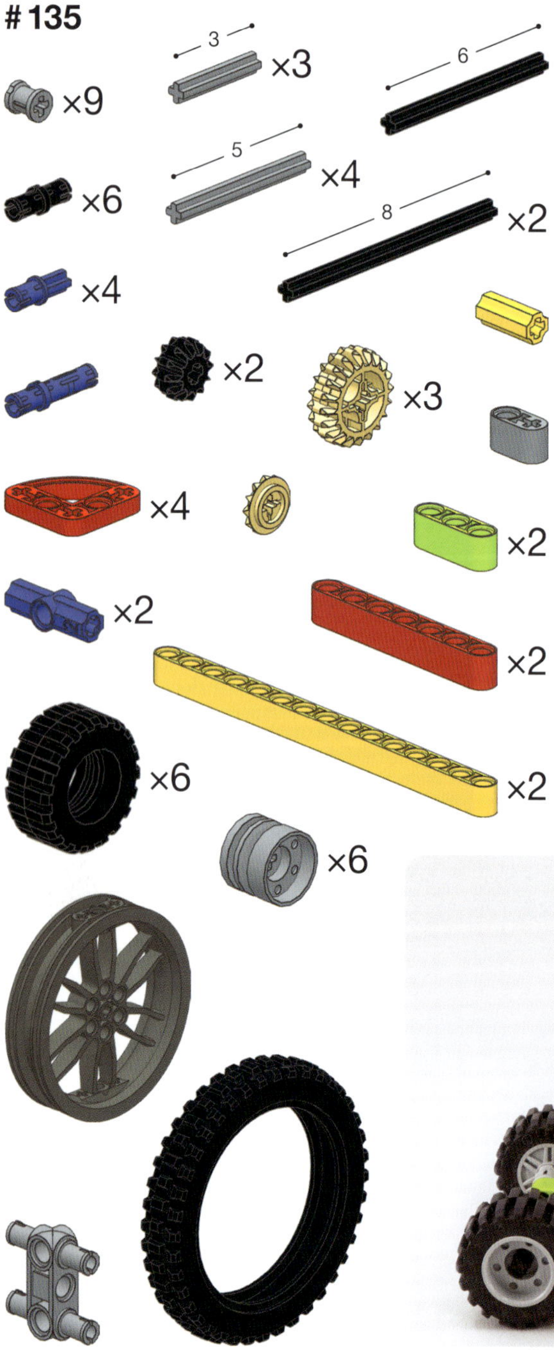

×9

— 3 — ×3

6 ×

5 ×4

8 ×2

×6

×4

×2

×3

×4

×2

×2

×2

×6

×6

The edge of a table

Push the car after setting these tires like this.

The car goes to the edge of the table and then backs up.

×4

×2

×7

×4

×4

×4

×6

×6

×2

3

4

5

8 ×3

×2

×2

×4

×3

×2

×2

×2

Wall (hold it steady so that it doesn't move).

Push the car after setting these tires like this.

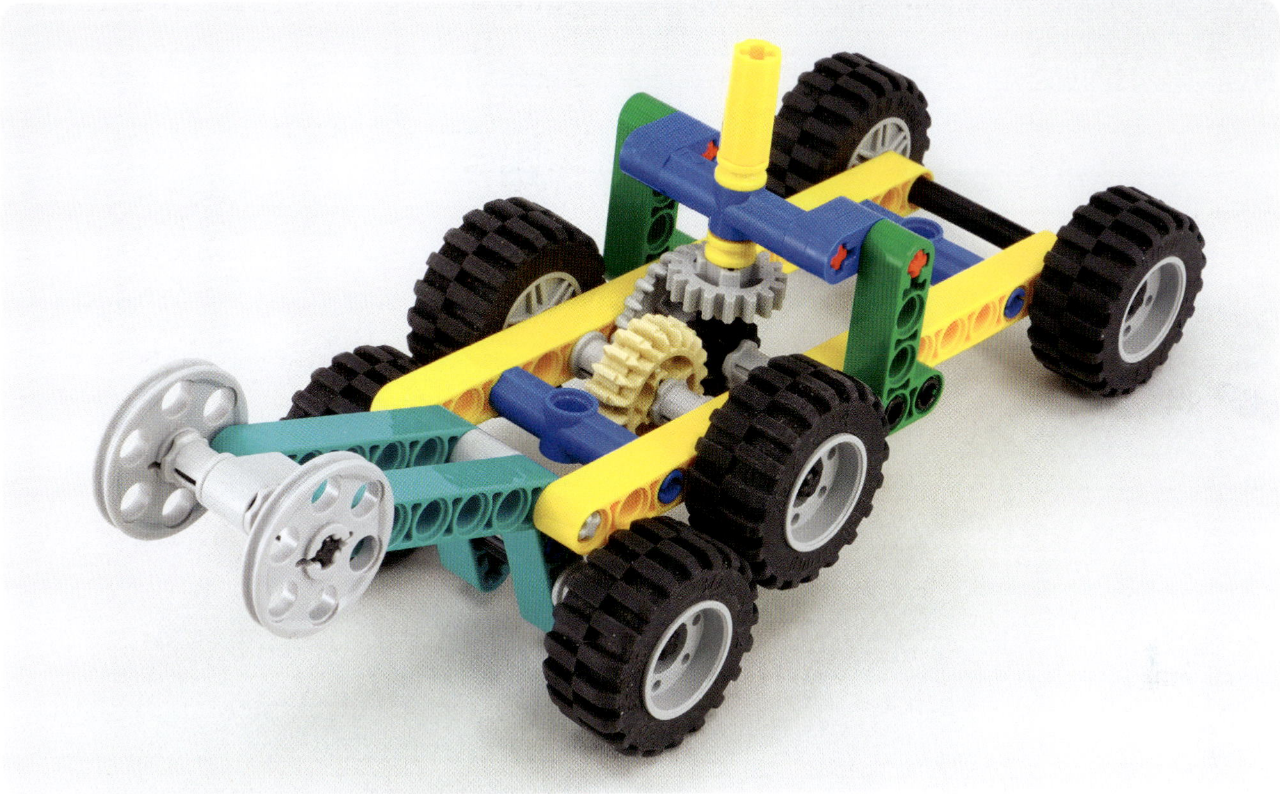

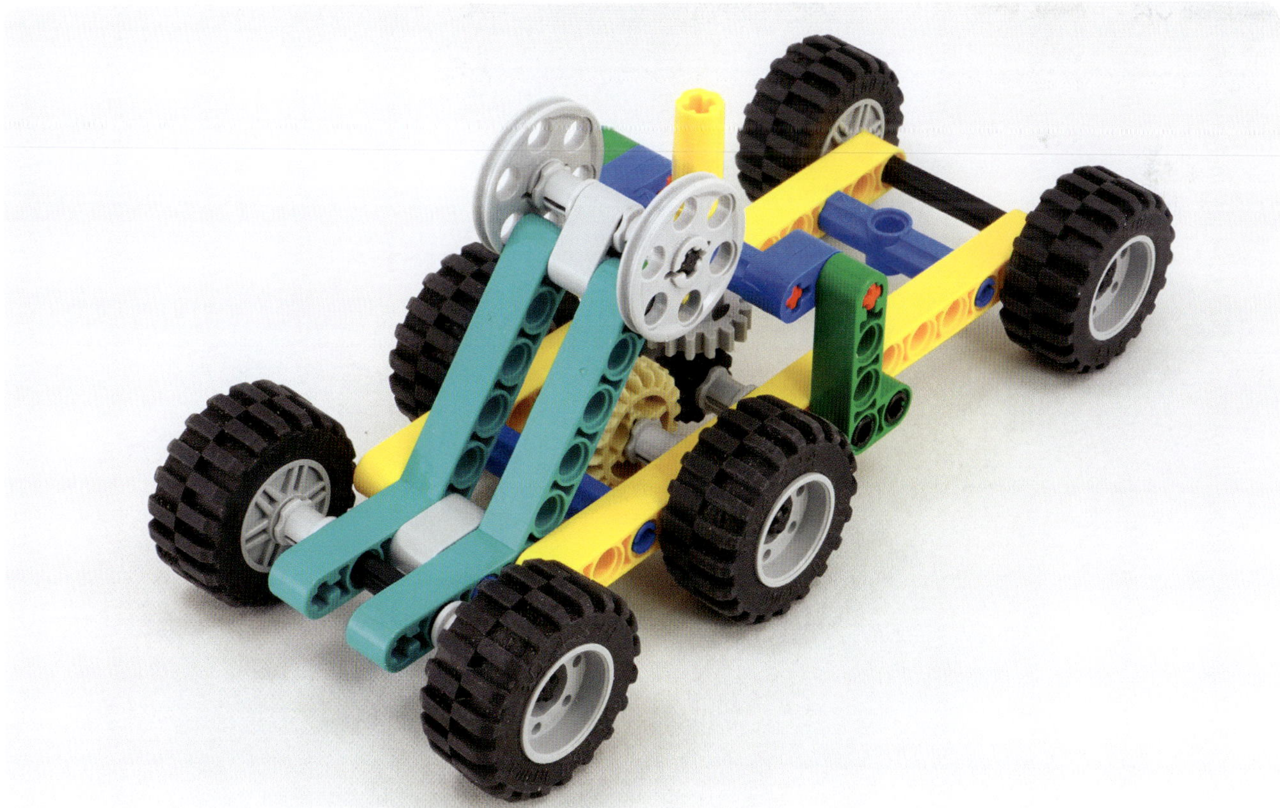

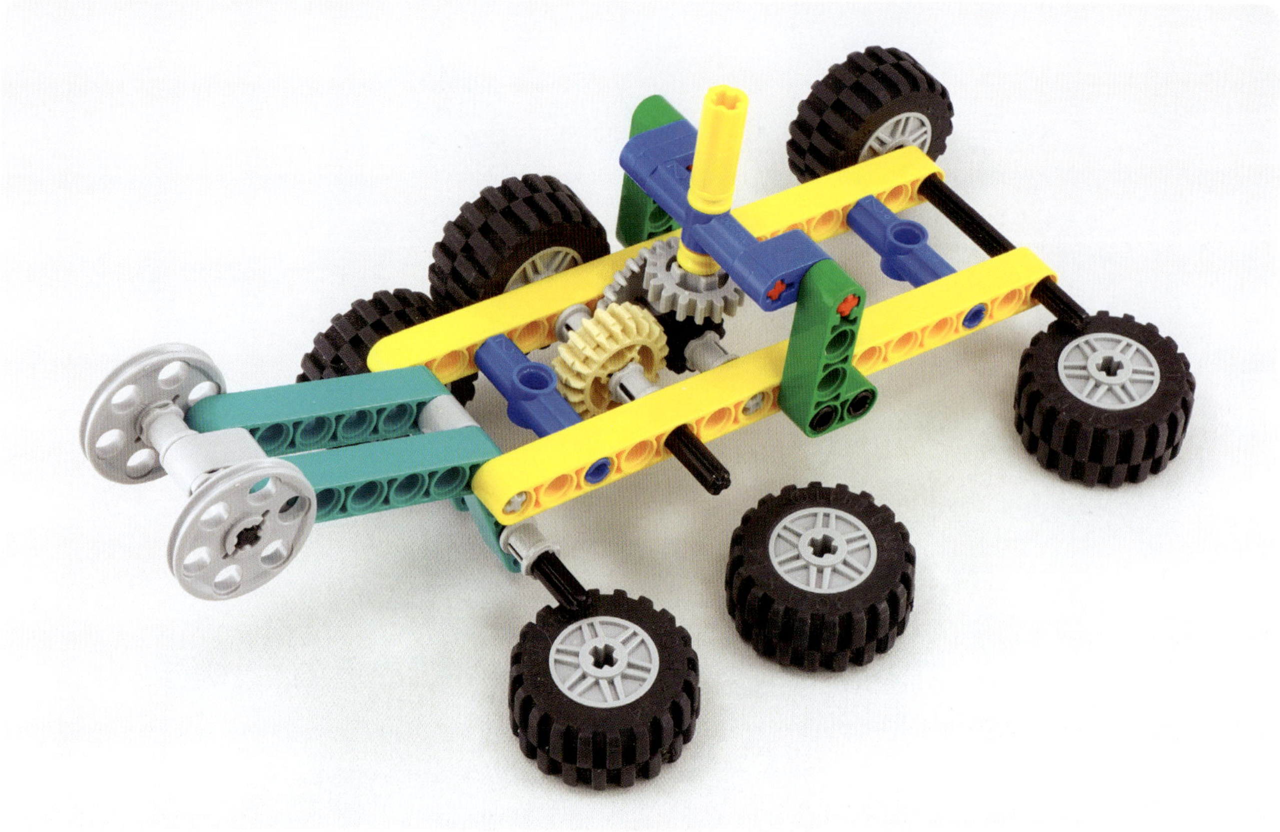

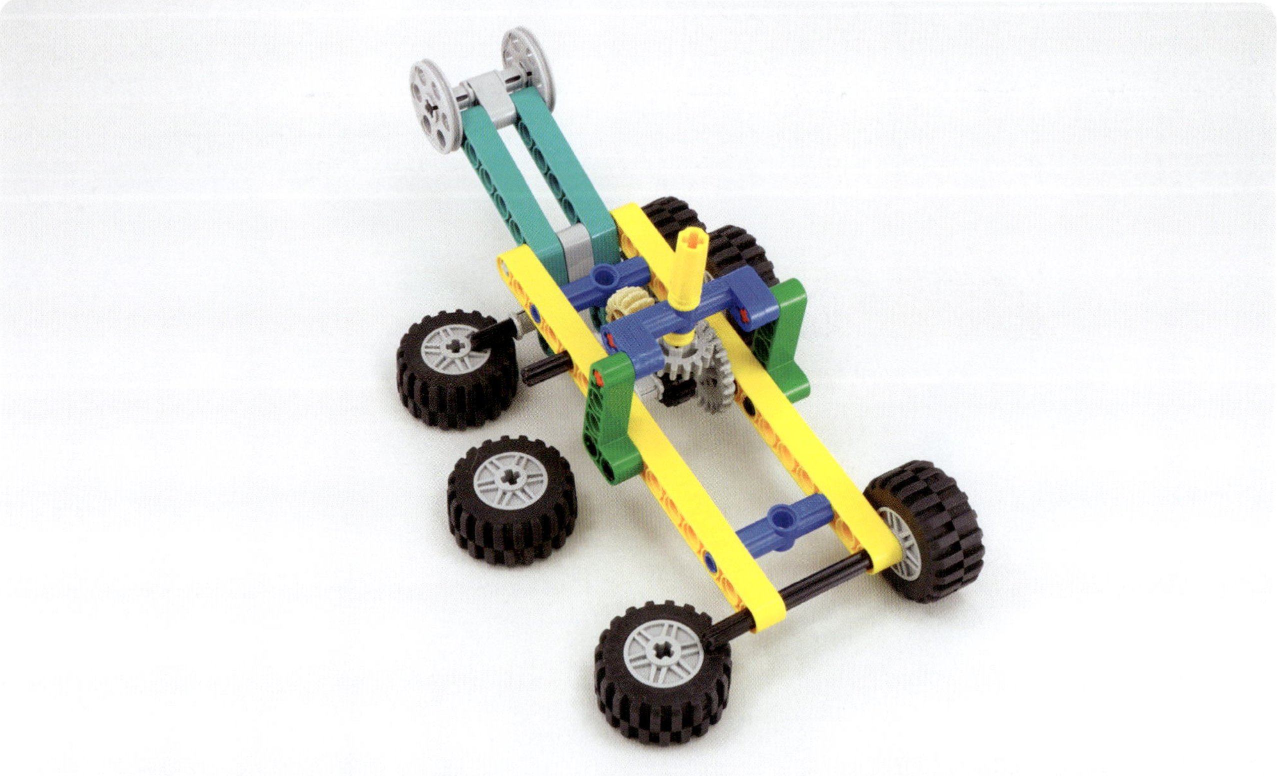

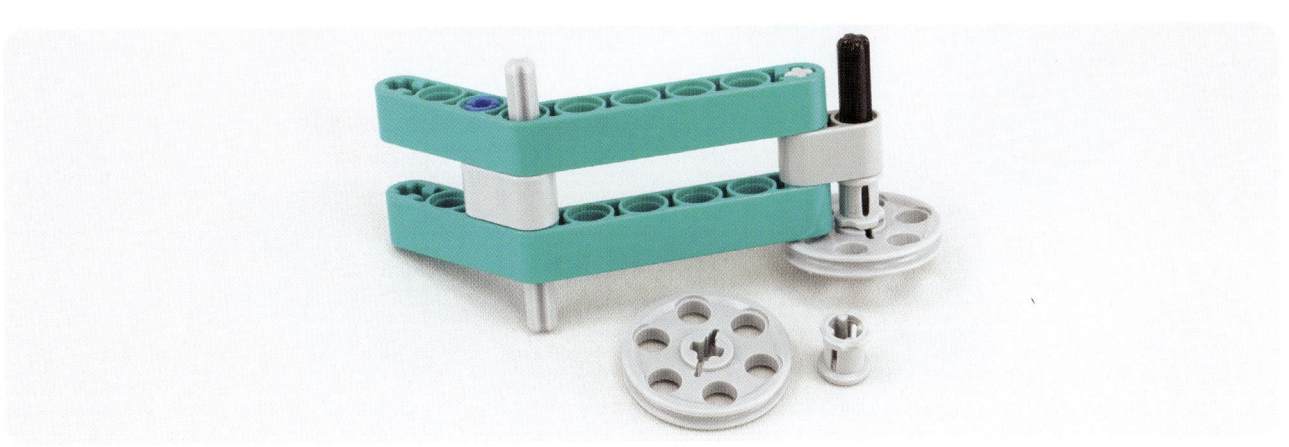

#137

×2
×2
×4
×12
×2
×2
×6
×2
×2
×2
×2
×2
×4
×4

3
4
4
5
8
10
12

Wall (hold it steady so that it doesn't move).

With this gear meshed with the one above it, push the car.

When the car hits the wall, this axis is pushed.

When this axis hits the opposite wall, the direction of travel changes again.

This gear meshes with the one above it and changes the car's direction.

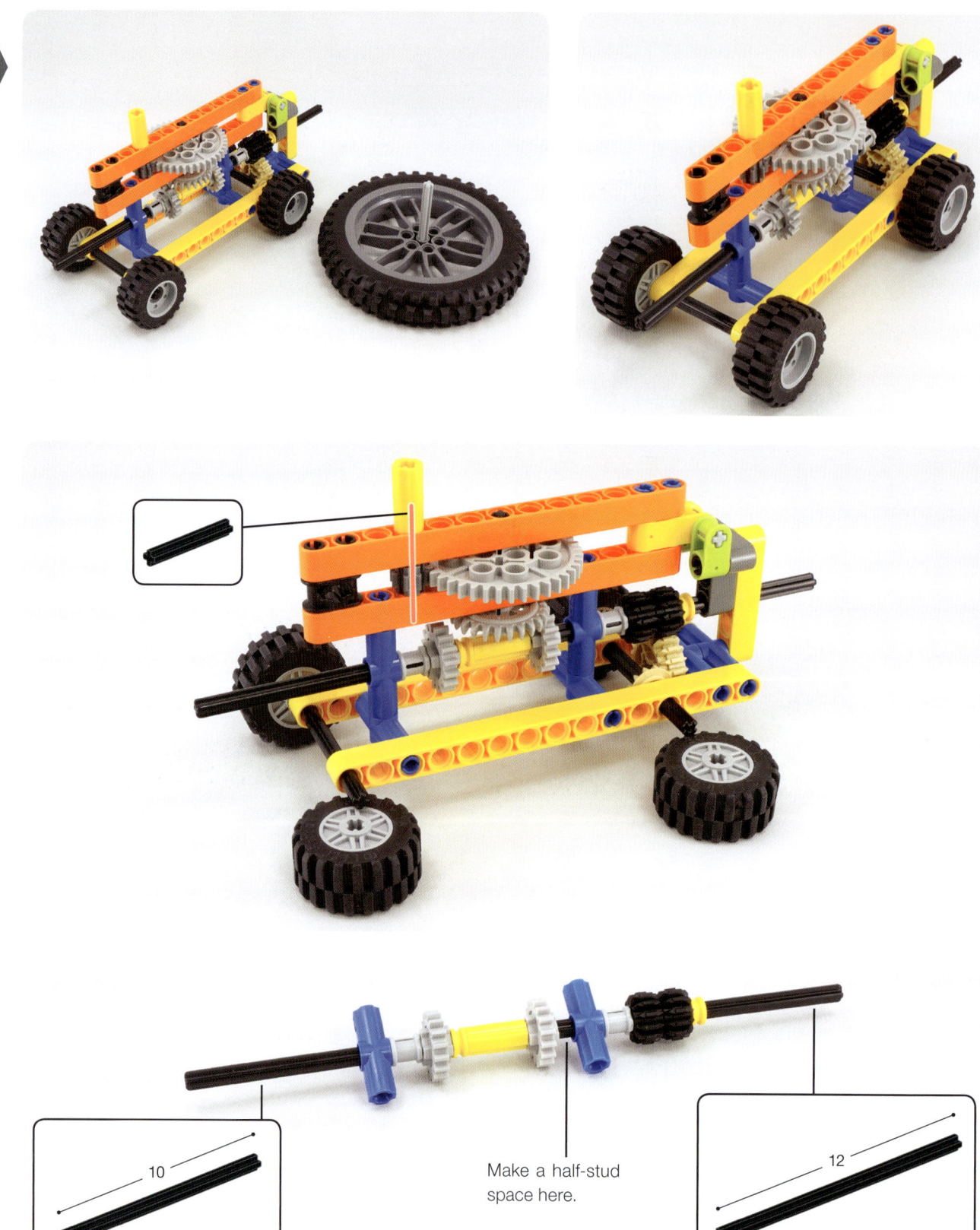

Make a half-stud
space here.

10

12

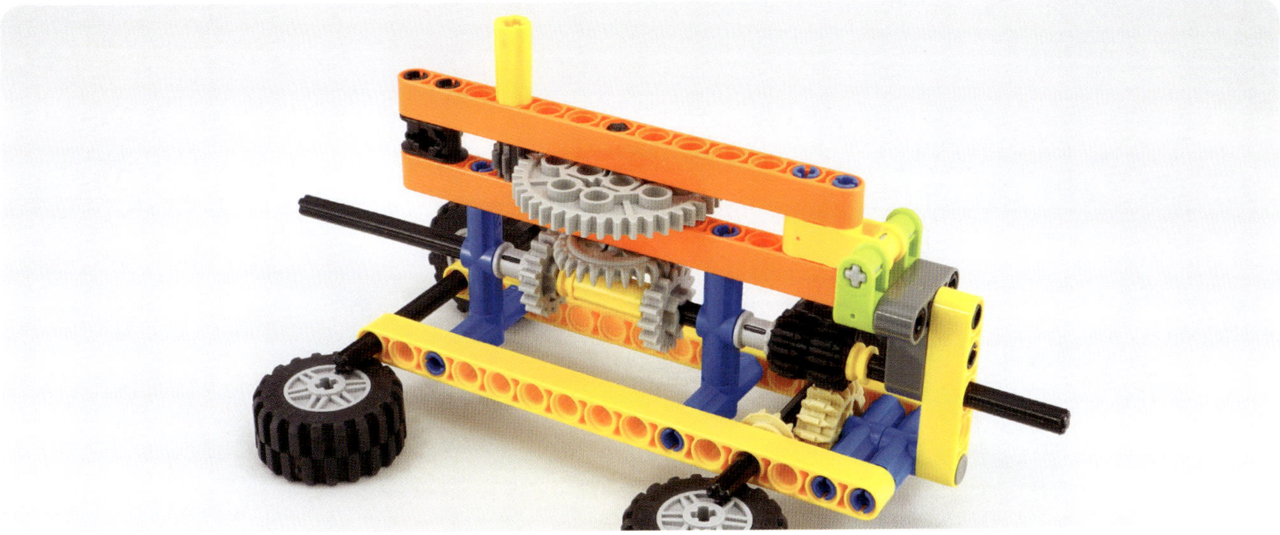

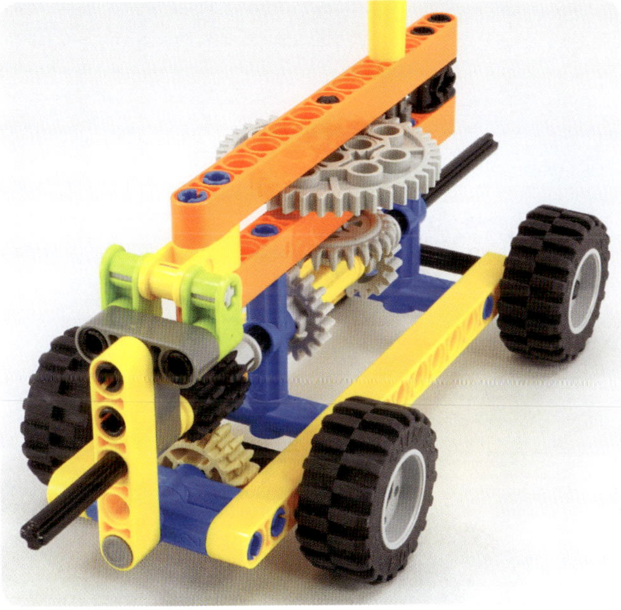

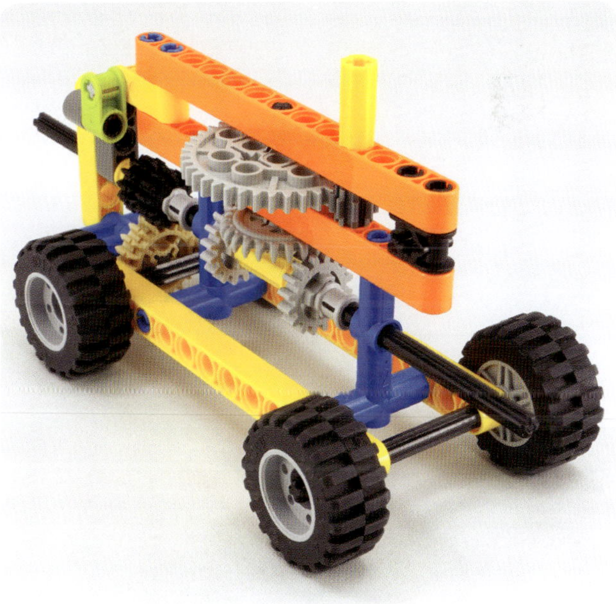

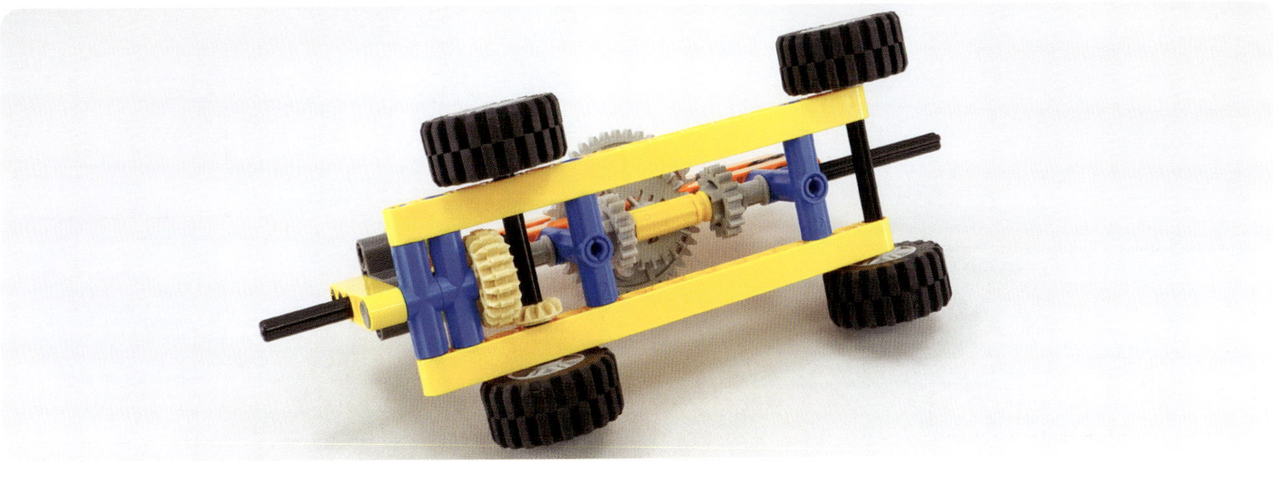

Other kinds of cars

#138

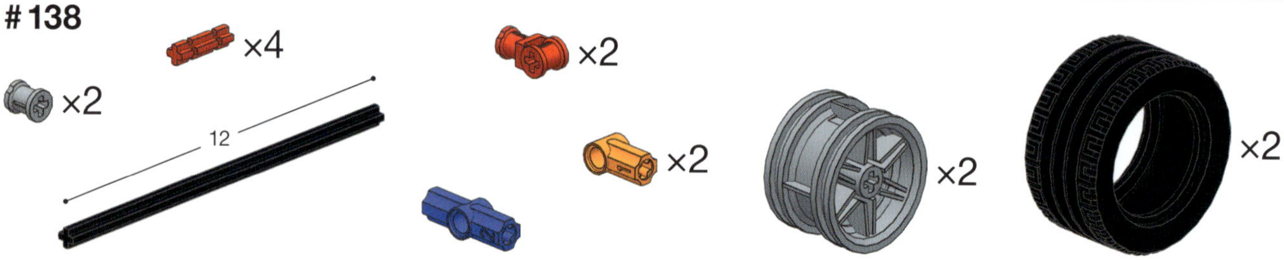

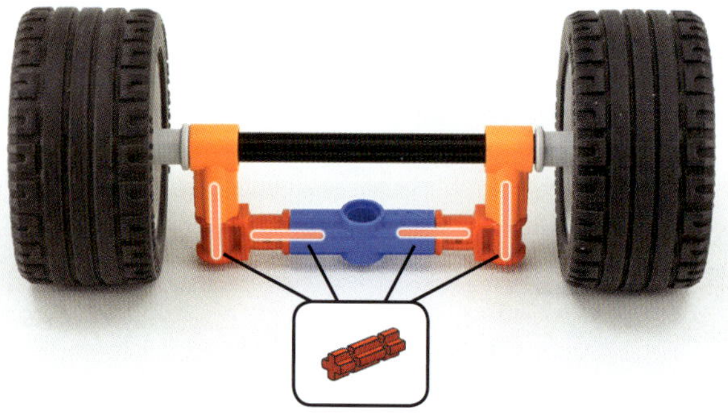

×4 ×2

×2

12

×2

×2 ×2 ×2

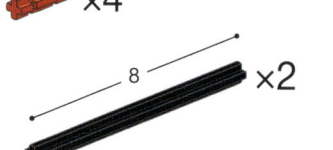 ×4

8 ×2

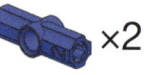 ×2

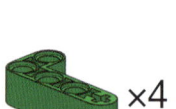 ×4

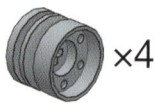 ×4

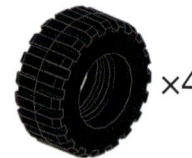 ×4

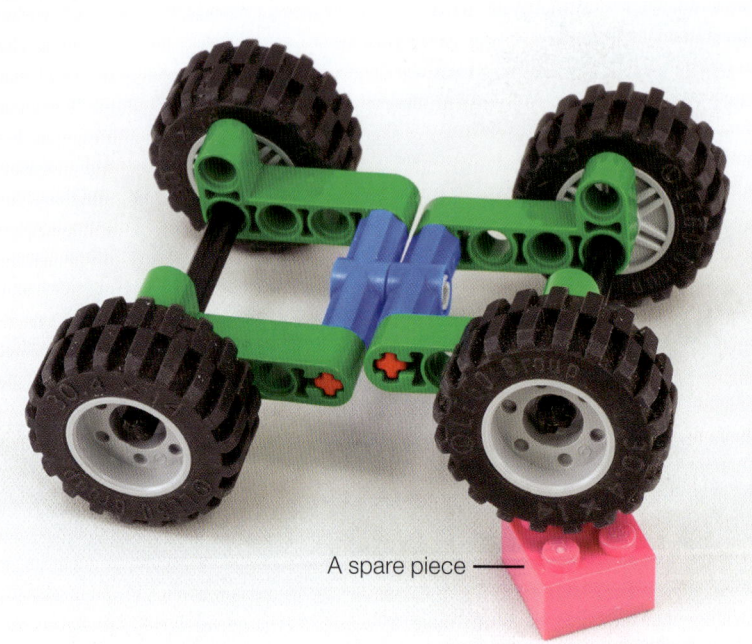

A spare piece ——

#140

#141

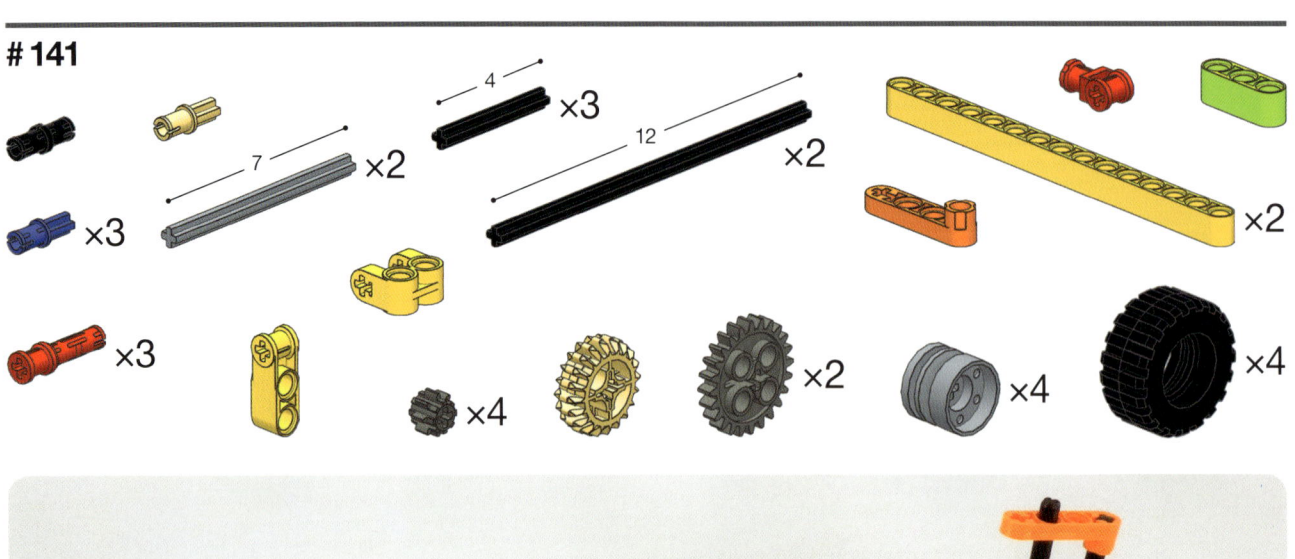

Flexing axles

This car is driven by the flexing of the two axles. This puts a lot of strain on some parts, and if you apply too much force, you could damage them. Be sure to pull the car back gently as you play.

Parts list

Part number

This is the maximum number of this part needed to build any single model in this book.

This is the maximum number of this part needed to build any single model in either volume of *LEGO Technic Non-Electric Models.*

#4265c x4 (x13)

#3713 x9 (x9)

#2780 x16 (x20)

#3673 x6 (x14)

#43093 x12 (x12)

#3749 x4 (x4)

#6558 x5 (x7)

#32556 x3 (x5)

#11214 x2 (x2)

#18651 x2 (x2)

#32002 x0 (x2)

#4274 x4 (x12)

#6628 x3 (x3)

#2736 x2 (x3)

#32062 x6 (x10)

3 #4519 x5 (x7)

#6587 x2 (x4)

3 #24316 x1 (x1)

4 #3705 x4 (x5)

4 #87083 x5 (x5)

5 #32073 x4 (x5)

6 #3706 x2 (x8)

7 #44294 x5 (x5)

8 #3707 x3 (x6)

9 #60485 x1 (x3)

10 #3737 x2 (x3)

12 #3708 x2 (x7)

#10928 x4 (x4)

#6589 x2 (x3)

#32270 x4 (x5)

#94925 x3 (x3)

#32198 x0 (x2)

#32269 x3 (x4)

#87407 x0 (x1)

#3650b x1 (x1)

#3648 x3 (x3)

#46372 x1 (x1)

#32498 x1 (x1)

#3649 x2 (x3)

#99009c01 x1 (x1)

#48452cx1 x1 (x1)

#18939c01 x1 (x1)

#3743 x2 (x5)

#3711 x45 (x45)

#2815 x6 (x6)

#6538c x2 (x9)

#87761 x1 (x1)

#3873 x31 (x31)

#55982 x6 (x6)

#18654 x1 (x1)

#6630 x1 (x1)

#57518 x21 (x21)

#89201 or #30648 x4 (x4)

#62462 x2 (x2)

#4716 or #32905 x2 (x2)

#62520c01 x2 (x2)

#92402 or #30391 x6 (x6)

#32054 x4 (x4)

#27938 x1 (x1)

#2739a x2 (x2)

#44 x2 (x2)

#32072 x2 (x6)

#92693c01 x0 (x1)

#56145 x4 (x4)

#32039 x4 (x4)

#6575 x2 (x2)

#6553 x0 (x2)

#4185 x6 (x0)

#x71 x1 (x3)

#44309 x4 (x4)

#32013 x6 (x6)

#98585 x0 (x2)

#x37 x2 (x2)

#32034 x6 (x6)

#731c05 x0 (x1)

#x90 x1 (x1)

#88517 x1 (x1)

#32016 x2 (x6)

#32192 x0 (x8)

#99012 x0 (x2)

#57519 x2 (x2)

#11957 x1 (x1)

#32015 x2 (x12)

#32014 x2 (x8)

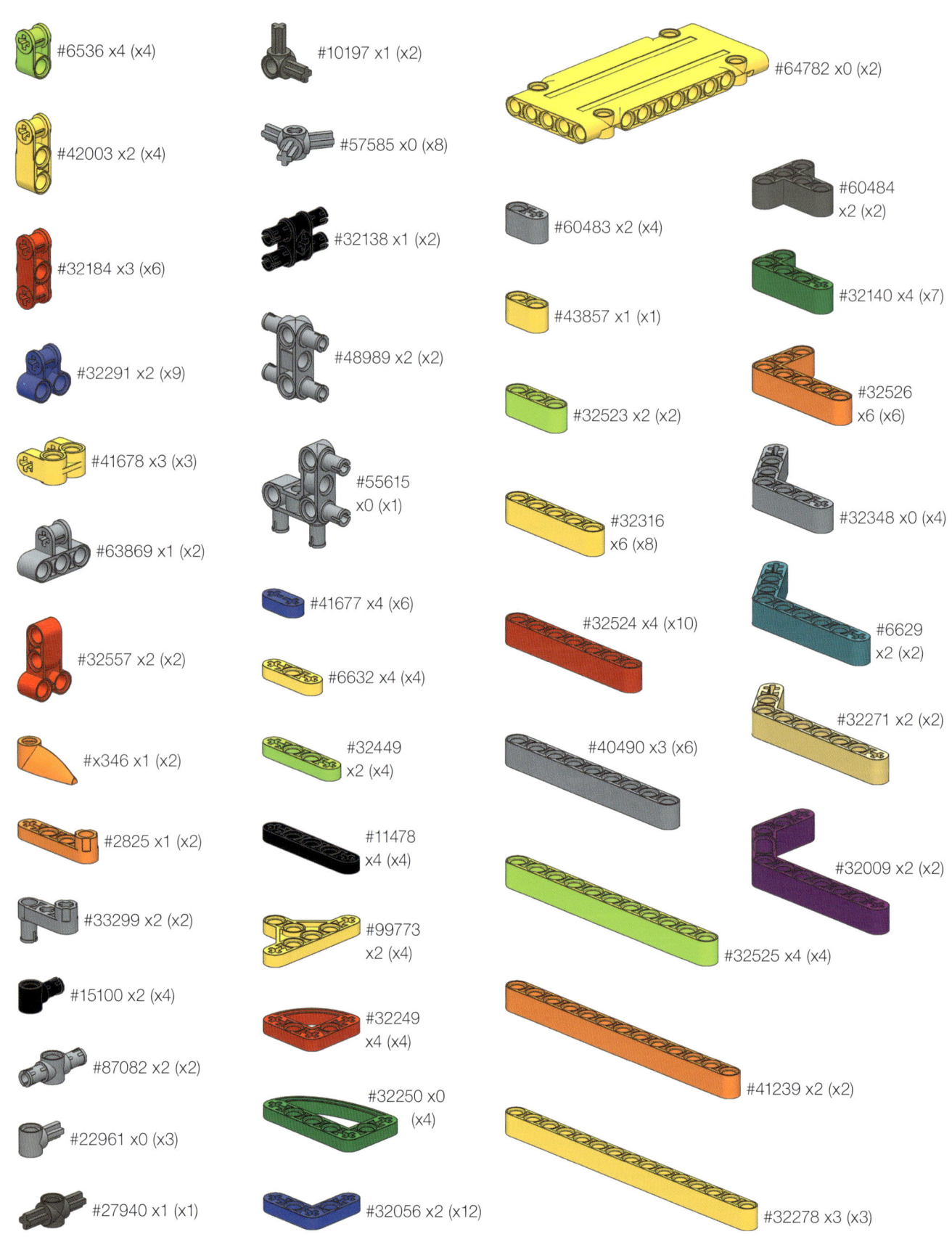

#6536 x4 (x4)

#42003 x2 (x4)

#32184 x3 (x6)

#32291 x2 (x9)

#41678 x3 (x3)

#63869 x1 (x2)

#32557 x2 (x2)

#x346 x1 (x2)

#2825 x1 (x2)

#33299 x2 (x2)

#15100 x2 (x4)

#87082 x2 (x2)

#22961 x0 (x3)

#27940 x1 (x1)

#10197 x1 (x2)

#57585 x0 (x8)

#32138 x1 (x2)

#48989 x2 (x2)

#55615 x0 (x1)

#41677 x4 (x6)

#6632 x4 (x4)

#32449 x2 (x4)

#11478 x4 (x4)

#99773 x2 (x4)

#32249 x4 (x4)

#32250 x0 (x4)

#32056 x2 (x12)

#64782 x0 (x2)

#60483 x2 (x4)

#43857 x1 (x1)

#32523 x2 (x2)

#32316 x6 (x8)

#32524 x4 (x10)

#40490 x3 (x6)

#32525 x4 (x4)

#60484 x2 (x2)

#32140 x4 (x7)

#32526 x6 (x6)

#32348 x0 (x4)

#6629 x2 (x2)

#32271 x2 (x2)

#32009 x2 (x2)

#41239 x2 (x2)

#32278 x3 (x3)

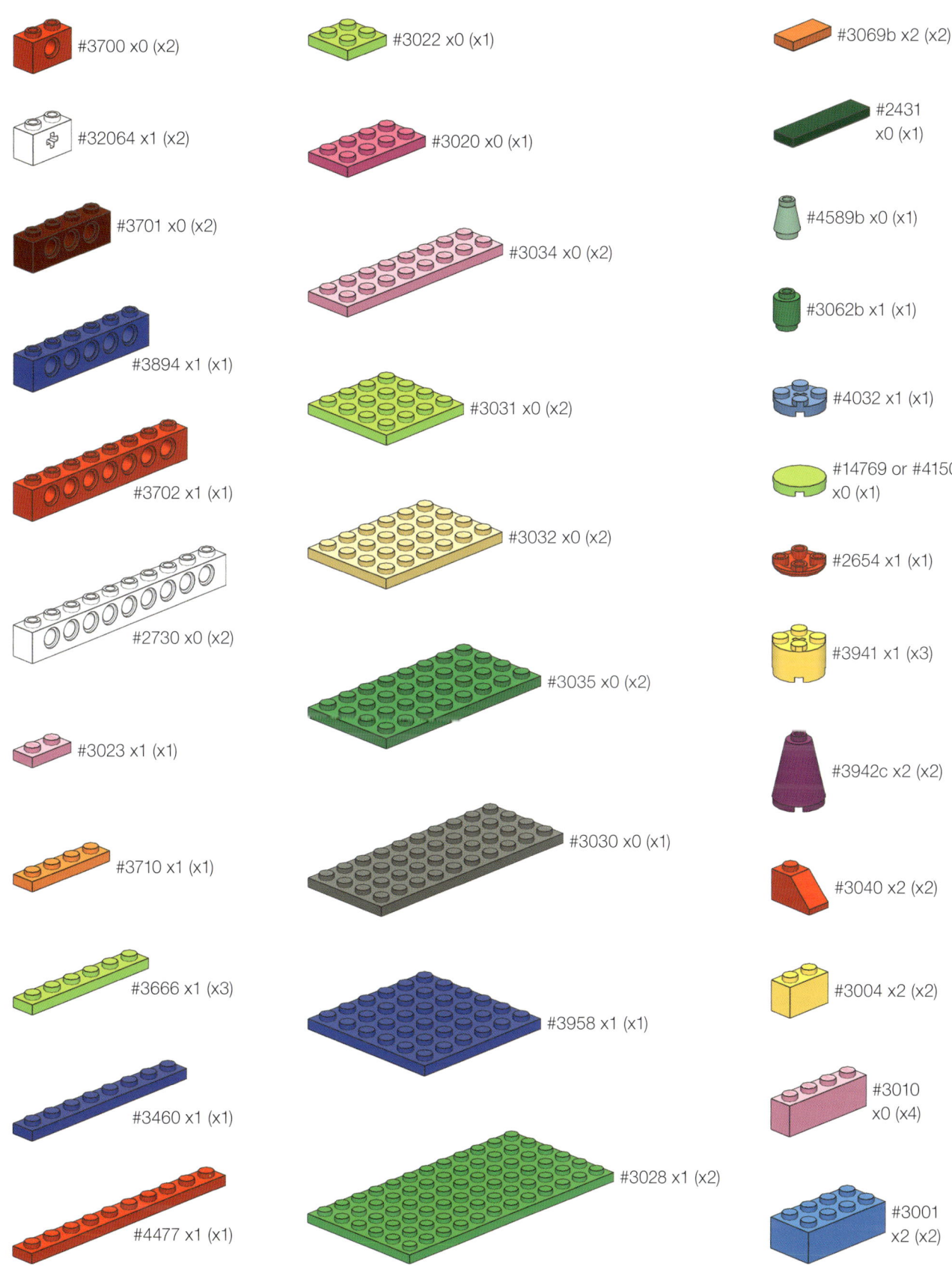

#3700 x0 (x2)

#32064 x1 (x2)

#3701 x0 (x2)

#3894 x1 (x1)

#3702 x1 (x1)

#2730 x0 (x2)

#3023 x1 (x1)

#3710 x1 (x1)

#3666 x1 (x3)

#3460 x1 (x1)

#4477 x1 (x1)

#3022 x0 (x1)

#3020 x0 (x1)

#3034 x0 (x2)

#3031 x0 (x2)

#3032 x0 (x2)

#3035 x0 (x2)

#3030 x0 (x1)

#3958 x1 (x1)

#3028 x1 (x2)

#3069b x2 (x2)

#2431 x0 (x1)

#4589b x0 (x1)

#3062b x1 (x1)

#4032 x1 (x1)

#14769 or #4150 x0 (x1)

#2654 x1 (x1)

#3941 x1 (x3)

#3942c x2 (x2)

#3040 x2 (x2)

#3004 x2 (x2)

#3010 x0 (x4)

#3001 x2 (x2)

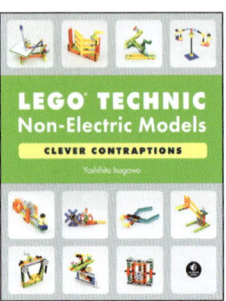
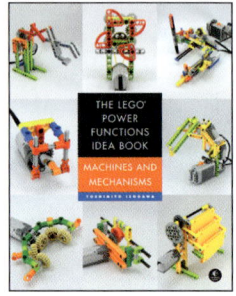
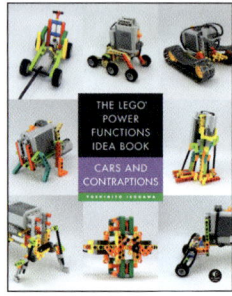
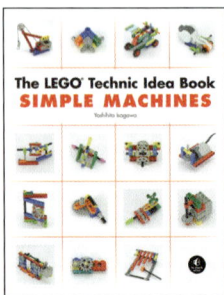
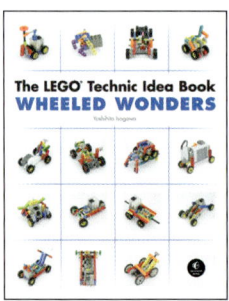
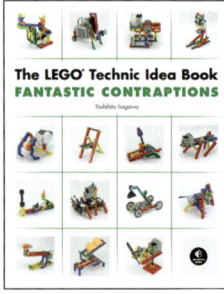